BADGE

A FIGHT

UNDER

FOR TRUTH

FIRE

CPL. CHAD JENSEN (RET.)

Ballast Books, LLC
www.ballastbooks.com

Copyright © 2026 by Chad Jensen

All rights reserved. No part of this book may be reproduced in any form or by any electronic or mechanical means, including information storage and retrieval systems, without permission in writing from the publisher, except by reviewers, who may quote brief passages in a review.

Hardcover: 978-1-966786-82-5
Paperback: 978-1-966786-81-8

Printed in the United States of America

Published by Ballast Books
www.ballastbooks.com

For more information, bulk orders, appearances, or speaking requests, please email: info@ballastbooks.com

The events described in this book are drawn from personal recollection and experience. Some names and identifying characteristics have been changed to protect privacy.

This book is dedicated to my wife, Amiee, without whom I would never have been able to survive this journey. Her unwavering support and love have been the rock from which I draw my strength.

TABLE OF CONTENTS

Introduction i
Chapter One: A Night at the Fair 1
Chapter Two: Growing Up SoCal 15
Chapter Three: Living My Dream 25
Chapter Four: The Worst Day 55
Chapter Five: Guilty Until Proven Innocent 71
Chapter Six: The First Trial 83
Chapter Seven: What We Learned 103
Chapter Eight: The Second Trial 117
Chapter Nine: Exoneration to Isolation 147
Chapter Ten: Into the Sunset 173
Chapter Eleven: Tay-Has! 191
Acknowledgments 197
About the Author 199
Endnotes 201

INTRODUCTION

Three years of a corrupt FBI investigation and evidence concealment, lying under oath, a fraudulent federal grand jury indictment, two federal grand juries, two federal grand jury trials, and multiple tests of faith. Not the way I had hoped and expected to wrap up a twenty-four-year law enforcement career.

A jury ultimately exonerated me of all charges. That should have been a good thing, but following the would-be redemption, I was summarily ostracized by my own department's FBI-friendly police chief and treated as if I wore a scarlet letter, leading to wrongful denial of promotion, overtime, and advancement in my career.

It wasn't supposed to be that way. I had visions of *Dragnet*, *Adam-12*, and *The Rookies* the day I decided to be a police officer. I was about eight years old, skateboarding with my best friend, Nick, down Highland Avenue in Monrovia, California, at the base of the San Gabriel Mountains. Like most kids that age, we were full of piss and vinegar but not an abundance of common sense. Skateboards don't have brakes, so we flew straight through a stop sign on Foothill Boulevard, a very busy town thoroughfare, without slowing down to even spare a glance at traffic. We were foolish on all fronts, but fun trumps sensible any day when you're eight.

As luck would have it, a Monrovia police car happened to be near the intersection and saw our hijinks. He turned in behind us, gave the siren a quick yelp, and pulled us over at the curb. I still remember every moment like it was yesterday. There we were, two little kids scared out of our minds while a quintessential 1970s officer—dark uniform, thick caterpillar mustache bigger than his nameplate, six-shooter on his hip, mirrored sunglasses—walked up to us. His radio crackled as he questioned us. "Where do you live? What were you thinking?" Nick and I stammered out answers, hoping somehow to get out of the mess.

Then he said something that stuck with me forever: "I could take you both to jail for this."

We froze as he let that sink in before adding, "All I'm trying to do is make you think. You could have been killed." And then he just smiled and told us to go home and tell our parents. We never did, of course, but that moment had a huge impact on me. That day, I knew exactly what I wanted to do. I wanted to be a police officer.

All throughout my childhood years, I had always held that dream of being a police officer. Although I did not have any law enforcement in my immediate or extended family, I couldn't shake the desire to be a cop.

Fast-forward nearly a decade. I was seventeen, living in Monterey, California, a small beach town, with my mom, stepdad, and grandpa. My grandpa was visiting while my mom and stepdad were out of town, so he was in charge. At the time, I had a beat-up, rusted-out gray 1973 Volkswagen Super Beetle, a $300 car at best. One night, I picked up a couple of buddies, and we drove over to Seaside, the next town over, to check out a nudie bookstore, because, you know, we thought we were cool.

When we got there, I figured I couldn't park right in front of the place, so I left my car in the pizza parlor lot next door—completely ignoring the giant "Tow-Away Zone" sign. We went inside, but within twenty minutes, we were freaked out enough to leave. When we stepped outside, my car was gone. I couldn't believe a tow truck had shown up that fast. I found a pay phone and called the number on the sign. A police officer showed up within minutes (the tow truck was probably waiting around the corner for idiots like us) and told me my car had been towed; I'd have to get it back the next day. Our "big adventure" was rapidly flaming out.

We started walking the four miles back home. About halfway there, we spotted a Monterey police officer parked in his squad car near the Naval Postgraduate School. His windows were down, so we walked up and told him what had happened.

He glanced at us. "You boys at that bookstore over there?" Uh-oh. But instead of giving us a hard time, he told us to hop in—he'd give us a ride home. That was my first time in the back of a police car. I remember the cage between us and the officer, the flashing lights on the dash, and the radio buzzing with messages in what sounded like a different language. I didn't understand any of them, but it brought back my eight-year-old declaration that I would one day drive a police car like that.

When we got home, my grandpa loaned me the $120 to get my car out of the impound lot (I never paid him back). Shortly after, my mom got home, and you know there was no way I was going to tell her!

A few weeks later, I unfortunately got into a heated fight with my stepdad and ended up storming out the door, headed down Robinson Street to my buddy Tim's place over in Pacific Grove. The next day, he drove me back home to get some stuff. I thought I would just get in and get out and the whole thing would blow over, but inside, my mom said, "I'm sending you down to live with your uncle Larry in Riverside." Shit. Well, I didn't have much, and Larry wasn't a bad guy. My stepdad felt bad about the fight and gave me a salvaged old VW Rabbit to make amends. It was a beat-up junker, but I drove that junker all the way to Riverside in October to finish my senior year at Riverside Poly High. Uncle Larry had converted the workshop in their garage into a one-room bungalow—no heat and no air conditioning, but it was my own space. My uncle took the time to explain to me how responsibility and work ethic would be a welcome addition to my life. I quickly got a job at Del Taco working the walk-up window, and my journey to adulthood was underway.

Life wasn't terrible living at my uncle's. There was almost a sense of family normalcy having him, my aunt, and my cousins around. After graduating in 1985, my high school girlfriend, Stephanie, was forced to break up with me, and I was heartbroken. I emotionally couldn't stay in Riverside anymore, so I moved up to Provo, Utah, to live with my brother. I was starting over, trying to leave Southern California behind. About a month later, I got a phone call—Stephanie was pregnant and had been kicked out of her house. My brother told me, "Don't marry her. Take responsibility, be a father, but don't get married. You're way too young."

Of course, I did the opposite. I drove back to Riverside, brought her to Utah, and married her, and we spent five turbulent years together, during which we had two children, Blake—my eldest son—and Brittanie. Being a young, married father, I did my best to provide by working at Macey's Sak 'N Save grocery store, making a whole $3.25 an hr. After one year in Utah, it wasn't quite panning out for us, so we packed up and moved back to Riverside, where I landed a job at Food 4 Less in San Bernardino. Our marriage lasted about another year but ultimately ended in divorce.

My dream of becoming a police officer would have to wait. Between having kids, making too many relationship mistakes, and divorce, that path seemed impossible.

After about a year of being single, I met, then married my second wife, Kimberly. We had two boys together, Nick and Haden. During that time, the Riverside County Sheriff's Office offered an after-hours reserve officer program, and I jumped at the chance. Reserve officers are basically volunteers—you go through all the same training and tests, but you don't get paid. I completed the year-long extended reserve training and became a reserve officer for Riverside PD—and I loved every minute of it. But after leaving the grocery business in 1995 and becoming an over-the-road driver of an 18-wheeler for Frito-Lay in Southern California, I had to quit the Riverside reserves to focus on my new gig and support my new family. Nevertheless,

Introduction

I couldn't get my dream out of my head, and I was determined to make it happen, whatever it took. At that point in my life, though, I didn't know if becoming a full-time police officer was the right decision. I had to be realistic with myself. I was on my second marriage with four kids. I had a good job, great pay, and stock options with Frito-Lay. For me to leave that job, I needed to be sure this was what I wanted, so joining the reserves again was a great way to find out if the police force would be a good fit.

I talked to an officer friend from my time working at the Pomona Food 4 Less who suggested I join the Pomona PD reserve unit. So, in 1996, I made the switch to Pomona reserve officer. I eventually applied for a full-time position with the PD, but they told me they didn't have the funds to hire me. Wanting to get started in my law enforcement career, I applied with the LAPD. I aced the written test, went through the oral interview, and scored high enough to be pushed through, but I still wasn't selected due to that old thing called affirmative action. For those of you not well-versed in California's affirmative action program in the nineties, let me illustrate it for you. When I went to the Devonshire Division Station to take my written test for police recruits, I turned in my test answer sheet to a basket marked "White." There were three other baskets marked "Black," "Hispanic," and "Other Races." The tests were scored on a curve with added points for any race other than white. During my oral panel interview at LAPD, the interviewer flat out told me my score of 99 didn't counter the added points that minority applicants got added to their scores. It was a harsh reality, but it was the reality in those days. I applied at several other departments in Southern California over the next year and eventually got a call from the Pomona PD lieutenant asking if I was still interested in a full-time police officer position. I was ecstatic, but the lieutenant told me I had to interview with the chief of police that day! At the time, I was driving my Frito-Lay 18-wheeler back from Boyle Heights and wearing my "potato chips" outfit—a Frito-Lay polo and tan shorts. Way to make a first impression.

The lieutenant introduced me to the chief, a large in stature, intimidating guy with baseball-mitt hands, who met me and said, "We've had a candidate withdraw, and I'm considering you for the position." The interview went well, and they offered me the police officer trainee position. Now, I felt my dream starting to take shape, and I was totally stoked.

At age thirty, I parked my truck for the last time, and four weeks later, I was part of the Rio Hondo Police Academy Class 140 in Whittier, California, with two other Pomona recruits. We graduated from the six-month program and became full-time officers in December of 1997. I spent five years on patrol, eight on motors (motorcycle patrol), eight on SWAT, and was promoted to corporal in 2010. During the early years of my tenure with Pomona PD, Kimberly and I divorced, and in 2001, I met my current wife, Amiee, who was an EMT in Pomona. We had two children, Ryder and Sydney, together.

Eighteen years after meeting Amiee, I was in the homestretch of a twenty-four-year career. But one summer night during an overtime shift at the Los Angeles County Fair, everything went to hell. A routine drunk-in-public arrest that made the news spiraled into a four-year nightmare when an opportunistic former-prosecutor-turned-defense-attorney saw a chance to make headlines. I spent the next two years in federal court, essentially fighting for my life against the FBI and Department of Justice (DOJ), who created a crime to fit their narrative and political aspirations.

Looking back, I never saw this coming. From the moment I stood on a Monrovia street corner being reprimanded by an old-school cop at eight years old, I thought police work would be like Mayberry. It wasn't. In reality, it was a drug-addled thug crashing through my front window, a wannabe Samurai coming at me with a sword, and a fifteen-year-old's life leaking away from a razor blade cut to the throat right in front of me. It was feeling scrutinized at every moment. I went from doing my dream job, no matter how hard it was, to being investigated, indicted, arrested, booked, and maliciously persecuted by people who I thought would be on my side.

This story highlights and exposes the absolute depth of concealment and levels of corruption the FBI and DOJ will go to. I was the key subject in a federal investigation riddled with evidence concealment and perjury, and although I (barely) avoided going to prison, the experience effectively cut my career short, filling my "retirement" instead with seemingly infinite civil litigation. This story exposes the darker side of federal law enforcement and how I became a target of institutional corruption and unchecked power.

There were many things that I was very proud of and that brought me joy in my career. I was directly involved in helping save lives, instituted a Recruitment Mentorship Program, and was instrumental in starting the department's Cops 4 Kids program. I was decorated many times for on-duty service, receiving accolades such as a Lifesaving Award, Officer of the Year, and several medals for locating stolen vehicles, as well as recognition for community events, significant arrests, and marksmanship. I remain immensely proud of what I accomplished and the opportunity to serve.

Yeah, I wish I hadn't had to write this book, given the trauma and life detritus that came with the events at the fair that night, but as much as I loved being a police officer and the city I worked in, I need to tell this story, to let the public know that not everything is always as it seems.

CHAPTER ONE

A NIGHT AT THE FAIR

Cotton Candy and Handcuffs

The Los Angeles County Fair doesn't exactly exude the typical county fair Americana. Traditional middle-America county fairs pulse with the energy of little kids pasted with ketchup and ice cream stains, often latched inside overheated strollers and wailing hysterically. Aged, gray couples sit on dirty metal benches, watching it all with faces etched in memories of their own heady days of youth. Among it all is rainbow neon, the tempestuous aroma of funnel cakes, hazy clouds from smoking barbecue, and the comforting rotation of the Ferris wheel.

The LA County Fair is different. Held annually in September at the Fairplex venue in Pomona (which hosts dozens of events each year, requiring ample police presence), it's usually scorching hot—north of 100 degrees is a common forecast—and like the Wild West out there, especially at night, and especially at the carnival section on Broadway. Similar to most fairs, specific sections are spread out all over—the petting zoo here, the food vendors there, Kiddie Corner on the next block—but the carnival attracts the knuckleheads and is typically where all the trouble happens. Alcohol-fueled, wannabe gangsters claim turf around the carnival or roam about instigating various states of chaos. The whole area smells of spilled beer on hot pavement, and general mayhem reigns. The best way to stay safe in this part of the fair? Stay out of it.

On one particular night in 2015, I walked right into it with my partner for that shift, Prince Hutchinson, both of us outfitted in Pomona Police Department uniforms. I couldn't have known it at the time, but that overtime gig at the fair would be my most infamous in a two-decade career wearing a badge. I couldn't have known that two years later, I would be arrested, booked, charged, held in waist chains and leg irons, and chained to a stool in a four-by-four room for seven hours in the Edward R. Roybal Federal Building detention center in downtown Los Angeles.

A last-minute schedule change had put me at the fair. Someone had needed help covering a shift, and I'd grabbed it. I had already been on shift at the fair for twelve hours, but six hours at time-and-a-half? Sign me up. Walking around the carnival at the fairgrounds under a California sunset was usually a crapshoot. Could be that nothing happened, or everything did. That night was true to the script until closing time at 10 p.m., when our radios broadcast a 415 call— "Victor-11, we'll be out with a 415 on Broadway by the marquee, send me a couple units." In any law enforcement environment, "officer needs assistance" calls like these initiate a great big world of shit hitting the fan right quick. In that case, it wasn't an emergency call, but in our department's history, we never had a surplus of resources for routine assistance calls, so a 415 meant someone really needed help and to get over there right now.

The majority of the law enforcement presence was placed in the carnival area where the call had come from, given that that area got the most rowdy. If there were fifty cops assigned to the fair on any given night, forty of them would be in the carnival section, while the remaining ten would cover the rest of the fairgrounds' 543 acres. On that night, I had already been working a long shift. I was hot, tired, hungry, sweaty, and my vest was sticking to me. My partner and I were on the north end of the carnival area, but the call had come from the south, so we headed that way, feet sore from the eighteen-hour day, thinking—hoping—it would be a quick on and off with the average end-of-day knucklehead causing problems. Prince and I arrived at the marquee area, packed with overserved patrons, as well as a mixture of nosey lookie-loos and families funneling down Broadway, trying to exit. Pretty much normal for this part of the fair. But things quickly started to spiral.

Two Hispanic males, father and uncle to that night's main character, Carlos Alvarez, had been detained for disorderly conduct after jumping a little white beer garden fence to pour themselves a drink after closing time. The attendant working the booth had told them, "We're closed, we're closed," but they didn't care. Instead, they'd hit her with a, "Fuck you, we're getting some beer," as they shuffled over to the taps. Fair security had intervened, calling the cops over, and the two subjects had been detained at the intersection of Budweiser Corner and a funnel cake stand on Broadway. All of this had happened while hundreds of fair attendees were exiting and the Thump Records booth bumped old-school hip-hop in the background, a compilation of Art Laboe's *Killer Oldies* and lowrider-esque music.

Due to the sometimes-violent history at the fair, the Fairplex contracted the Pomona police department to provide extra help and back up their private security. Even though the fairgrounds are on county property, they're privately owned, meaning people can be arrested for trespassing and public intoxication. The night had almost been over, and the two original officers were just trying to get through their shift. They'd told Alvarez's dad, "Hey, just go." But the dad had started up, attempting to bait them into a fight, saying, "I work at Arcadia [Methodist Hospital]. You're gonna end up like Diamond. Fuck you, pigs." Naturally, a crowd had gathered and started heckling the police. By that time, the officers had determined that the two offenders were also 647(f)—drunk in public—and decided to call for backup.

Officer Sean "Diamond" had been one of my SWAT teammates for several years. He'd also been a good friend. We'd had similar challenging pasts, came out on the higher side of things, and spent a lot of time together away from work with mutual friends. He'd taken over my breaching position after I left the team. A year later, he'd been killed in the line of duty—a 12-gauge slug to his head during the breach of a no-knock, dynamic entry warrant. They'd taken him to Arcadia Methodist Hospital, but he didn't make it six hours. And now this guy, Alvarez's dad, was out here telling our cops they were gonna end up like Diamond.

At that point, things were tense. The dad and uncle were big guys, obviously drunk, and there were just two cops with them until Prince and I arrived on scene. It was around 10 p.m., and we'd been hoping to be cleared to go home soon until this all went down. Several other two-man units arrived as the officers escorted the now-handcuffed suspects northbound down Broadway, followed by a growing crowd of agitators, taunting and shouting at them. Prince and I tried to position ourselves behind the officers to cover their backs and keep the crowd at bay if things went south. In these situations, instigators in the crowd would often start throwing rocks, bottles, trash cans, or whatever they could get their hands on. As rear security, it was our job to ensure that didn't happen.

Most of the crowd slowed down when we told them to back off—except for one. Carlos Alvarez, a five-foot-eight, two-hundred-pound young man. As he was approaching, I put my hands out—"Back off. Slow down. Stop." He didn't listen; he just breezed right past me, fast-walking to catch up to his dad. I reached out for him, but he was just outside of my grasp.

I called out, "Hey, can you guys step over here for a minute?"

He responded, "Nah, man, I'm going to the front," and continued to accelerate, closing the distance to the officers' backs quickly.

Now, I had a problem! There was nothing between him and the backs of the escorting officers. So I had no choice. Trying not to escalate the situation, I turned and caught up to him, taking hold of his arm; I was trying to keep it cool, trying to separate him from the crowd and simmer things down. I escorted him, now passively resisting and dragging his feet like a nine-year-old who'd just had his Xbox taken away, into a small alcove near the grandstand entrance. It was even darker in there, with a green-and-white wooden fence enclosing part of the space, and for a brief moment, things seemed to calm. But as I started to check him for weapons, he suddenly tensed up and flinched, an indicator of fight or flight.

And just like that, the fight was on.

He pushed off the fence, spinning us around, and the next thing I knew, a fist was flying past my face. I cross-checked him with my right forearm in the upper chest, trying to create space and get him off me. Instead of retreating, he grabbed my shirt and pulled me toward him. Typically, when someone fights a cop, they try to escape. Alvarez wasn't trying to run—he was pulling me in. I cross-checked him again, trying to get distance between us; then Prince came in and bear-hugged him from behind, attempting to yank him back. Another officer arrived with a baton—cracking Alvarez on the foot at the bottom of his shoe in an effort to bring him down.

Prince took him down, but Alvarez wouldn't stop fighting. He was on all fours, resisting. More officers arrived, including a sergeant, who removed the probes from his Taser and arced it in the air—letting Alvarez know he was about to be tased if he didn't stop.

Alvarez finally surrendered, flattening out, and was handcuffed. As they stood him up, he smirked and said, "Y'all hit like bitches." The crowd had now swelled even larger, buzzing with tension, and there were about a dozen officers in the area.

Throughout the struggle, I'd made a conscious decision: I wasn't going to punch him. Even though I'd had every right to, I knew how bad it would look. That's why I'd used the cross-check—to create distance without making it look dirty or aggressive and to, most importantly, get him off me.

Of course, someone had been filming the interaction, but the video didn't show the lead-up—the crowd, the warnings, the attempts to de-escalate—only the moment things had turned physical.

After Alvarez's arrest, we took him to Station Five, the small substation inside the fairgrounds. His dad, already in holding, started yelling, "Why is my son in here? He's a minor!" With his size, Alvarez didn't look like a minor, but the officers immediately removed him from the adult holding area and arranged for transport to the hospital—standard procedure anytime force is used. (To be clear, the only official use of force in this case wasn't our physical struggle; it was the baton strike.)

After receiving medical clearance, Alvarez was interviewed, photographed, cited, and released to his mother later that night from the hospital. Prince wrote the main report, I wrote up a supplemental report, and everybody went home. Just another summer night at the LA County Fair, one that reminded me that policing in the Midwest or on the East Coast was probably a completely different world compared to what we dealt with here. It was constant chaos, constant hostility. For a long time now, law enforcement had been vilified. Yeah, there were bad cops out there, but the sheer amount of bullshit we dealt with every day was on another level. On some late nights in our city, there were only six of us working the streets of Pomona. Six guys patrolling and protecting a twenty-six-square-mile town of 180,000. Pomona is known as a high-crime city with a diverse population of working-class, mostly Hispanic people. Our city is plagued by gang violence, homicides, and prostitution—Beverly Hills it ain't. Being a police officer there meant trial by fire and riding by the seat of your pants, and I wouldn't have had it any other way.

September 16, 2015, was the day of the fair incident, and since that routine day, I had continued on per usual, pulling my shifts and even working more overtime at the fair that year and the next. But on October 25, 2017, I was indicted on an abuse under color of authority (excessive use of force charge), deprivation of civil rights, and lying on a police report. The next day, I was arrested, photographed, booked, printed, berated, and chained to that stool in lockup at the Edward R. Roybal Federal Building in downtown Los Angeles. It was just the starting point of this horrible journey.

Internal Affairs and Channel 4 News at 11

The morning after the fair incident, two things happened: After all our reports were written, the internal affairs division was notified, and the sergeant in charge of the squad at the fair the night of the incident had to open up an administrative insight. Basically, an administrative insight is a patrol-level initial review of the incident, more of an internal check—we look at what happened, see if everything was handled correctly, and consider what we can learn from the situation.

Since this was an overtime shift fair incident, the sergeant in charge of the squad at the fair that night handled the administrative insight, while internal affairs ran their separate, lengthier investigation. The sergeant gathered all the reports and statements from everyone involved, then wrote a synopsis of the whole thing—what happened, the results—and sent it up the chain.

The next day, the sergeant emailed a few of the officers involved in the use of force but didn't CC me (he felt my report was sufficient). His email said, "Hey, we need to clean up some of these reports. These kinds of things can lead to litigation." By that, he meant civil lawsuits. He wanted us to be more specific about times and locations mentioned and make sure we got names correct. Including these particulars within the report makes it easier for the reader to understand the actual course of events that night and helps us document the facts in detail in case of further investigation. But this was an insignificant arrest and charge—148 PC, delaying or obstructing an officer, was a low-level misdemeanor that was rarely, if ever, criminally filed. In fact, it was usually dropped by the district attorney prior to arraignment. Moreover, there were no injuries. Drunk arrests like this happened multiple times a night in the carnival area at the fair. It was a zero on the radar.

So, everyone in that email thread made a few corrections here and there, and the sergeant attached the "Hobson video"—a cellphone video taken the night of the fair incident by a random bystander named Hobson. It was a grainy video of the fairgrounds at night with minimal lighting, so the video was not the clearest, but it did show some of what happened. The lieutenant, captain, deputy chief, and chief of police all reviewed it and added their two cents, and in the end, a week later, the admin insight came back—no issues. Everything was within policy, everyone signed off; it was a done deal.

Then, maybe a week or two later, internal affairs hit me up. Turns out Carlos Alvarez's mom had filed a complaint claiming excessive force and that her son had been wrongfully hurt at the fair. They called me in, and I didn't even bring an attorney because, really, what was there? I cross-checked the guy twice, didn't punch him, didn't kick him. This wasn't a Rodney King, wasn't a George Floyd incident. No use of force. No injuries. Nothing.

The IA sergeant running the interview, Mike Neaderbaomer, was known within the department as very by the book and well-versed in officers' rights. He was very adept at conducting thorough and fair investigations. Our career paths hadn't previously crossed—we'd always worked in opposite squads—but here we were. Mike ran through it step by step, methodically recreating everything that had occurred that night. After about two hours, we ended the interview, and I didn't really hear much of anything for a while. Once completed, the IA report then goes up the chain again, all the way to the chief, who reviews and signs off: "Use of force was justified and within policy." These types of internal affairs investigations, involving multiple witnesses and subjects, typically take nine to twelve months. This one followed that trajectory and ran its course, but it was never officially closed since the FBI opened an investigation into this incident. Anytime an outside agency opens an investigation regarding the same incident that internal affairs is investigating, the case is "tolled," meaning that the internal affairs investigation at the police department level is temporarily suspended, pending the outcome of the federal investigation. They do this so that any new information gleaned from the federal investigation can be included in the police department's own internal investigation.

In May of the following year, the juvenile 148 PC obstruction case was finally going to state trial, where a judge would decide if Alvarez should be charged or if the charges should be dropped. The night before the pretrial hearing, Channel 4 News just so happened to run a piece on "police violence at the fair." And to my surprise, that same Hobson video was plastered across the TV. It was still choppy and glitchy, but it was now completely lightened and a lot clearer. Of course, the news picked the perfect millisecond freeze-frame—Alvarez just standing there with his arms down and my elbow coming toward his solar plexus. That's what gets ratings.

My wife, Amiee, called me at work, saying, "I think you're on the news." Confused and a little curious as to how it would be spun by the media, I sat down and watched the recording of the news piece that she had recorded earlier in the evening.

I didn't think much of it, but the next day, I sat down with the deputy chief and asked, "Is this going to be problematic?"

He responded, "I saw the video last night, and when you mix it up with the cops at the fair, this is what happens. I'd defend this seven days out of seven. You did nothing wrong." I thought I could trust him to have my back, but I would learn that was a mistake. When it was comfortable, no problem, but the second it got not-so-pretty, he turned 180 degrees against me.

That night at the fair, I hadn't been trying to escalate anything. I wasn't out there looking for trouble. My whole approach was low-key—just separating Alvarez from the situation, trying to calm him down, and checking him for weapons. That was it. A quick waistband check, make sure he's not armed, and move on. But the second he grabbed me, everything changed. As a police officer, there's a difference between someone running away and someone grabbing you. That's when instincts take over. It hadn't appeared to be a life-or-death scenario, but it had still been a fight, and I'd known it could turn deadly at any second.

Unfortunately, we didn't have body cams back then. Our officers association had been pushing for them, but the department was dragging its feet. And of course, after this, they finally got them. If I'd had one that night, none of this would've even been a conversation.

People think cops go out looking for fights. That's not me. If anything, I talk too much. But when the talking's over, and it's time to go to work—you go to work and get the job done, whatever it takes.

State 148 Trial

Shortly before the Channel 4 News bit aired, I was still working in the training division doing backgrounds when I got a call from the district attorney. She asked me to come into her office to discuss the case, and when I arrived, she played an enhanced version of the video from the fair that night. I was honestly stunned. Up until that moment, I had only seen the dark, grainy version. I even had her play the video a few times, and with the enhancement, everything looked much clearer, like high noon rather than midnight. I had always been confident in my memory—I had told her before, as sure as I'm sitting here today, that it was his right hand that swung. But when I watched the video, I saw his left hand right up by my face.

We tried to pause on frames and analyze the footage, but at the end of the day, she'd shrugged it off: "Well, what you remember is what you remember."

May 15, 2016, was the pretrial hearing for Alvarez, the first day of the state trial. Again, the only people in the courtroom were those directly involved in the case. I took the stand and met Doug Gaines for the first time. Gaines was a typical younger, self-absorbed, smooth-talking LA-type defense attorney working for a law firm with the reputation of defending every dirtbag in Los Angeles County.

He started questioning me, going over my report and everything I remembered from that night. Then he said, "We have a second cellphone video for you to take a look at."

What! I had never seen this second video.

He played it, and I watched Carlos following his dad, trying to catch up. I glanced over at the district attorney, thinking, *Are you kidding me?* She'd never shown me this and just sat there, doing nothing. The right move would have been for her to stand up and pause the trial, but she stayed silent.

I was on the stand, stunned, trying to reconcile what I remembered with what I saw on this new video, without any context for distance, speed, or perspective. I struggled through it. The video appeared to be from the night in question, but I wasn't visible in the footage.

Gaines asked, "Is that your voice?"

I paused. "Well, it sounds like me."

He kept pressing: "Is he within arm's reach?"

"No, but he's getting there. If I don't do something, he will be."

Every word was picked apart. Gaines did his thing, creating doubt and confusion and making me question myself. After I was off the stand, Carlos testified, and the judge decided there was enough evidence to move forward. Gaines had done his best to get the case against Alvarez dismissed, saying that when the police had stopped Alvarez, they had not had cause to do so. In this preliminary hearing, the judge ruled against Gaines' motion to suppress and found that there was sufficient evidence that contact with the suspect had been warranted. The trial was officially on.

A week later, we returned for the 148 trial—a juvenile criminal trial with the same players. But this time, Gaines showed up rolling a gigantic sixty-inch TV into the courtroom and went to work with one goal: find any minor inconsistency with my testimony and turn it into a "lie."

Gaines kept pointing to the video. "What does the video show?" He was trying to blur the lines between my memory and the footage. We went in circles ad nauseam for hours of testimony.

At one point, he asked, "In the video, do you see him throwing a punch?"

From that angle, I didn't. "No, in the video he's not throwing a punch," I admitted. "But he's got what appears to be a flailing hand that comes right up to my face—you can see me react and stiffen up to avoid getting hit."

In the heat of the moment, you don't have time to pause and analyze. You don't get to say, "Hold on, let me rewind that in my brain." You react to what you see and feel in real time. Ultimately, Gaines twisted the question. "Now, as you sit here today, are you willing to admit that he never swung at you. Despite what your perception might have been that night, he never swung at you?" He was trying to separate what I'd perceived during the real event and what was on video. The only problem was that my viewpoint when trying to make an arrest that night had not been through a cellphone fifteen feet away, slowed to half the speed. I'd only had milliseconds to make a decision and react. At the time, with limited sight and no time, I'd thought he was throwing a punch.

Now, as I sat in court testifying in a sterile environment months later, looking at snippets of slowed-down video, one could reasonably agree that it could've been a flailing fist, not a punch.

Being rational and unafraid of the truth, I responded that it could've been a flailing hand and not a punch, bearing in mind that my perception at the time had been that a punch was being thrown in a fight. After closing arguments, the judge spoke a bit about a lack of preparation on the prosecution's side and confusion about what may have actually occurred that night at the fair. As with all criminal cases, if there is any reasonable doubt that the intent was not there or if certain elements of the crime are missing, a conviction is not possible; therefore, the case was dismissed. Carlos went his way, and I went mine, assuming that was the end of it.

About a month later, I heard rumblings that the FBI was snooping around the department. Initially, I was told it had nothing to do with me, that they were looking to put a consent decree on our department. Under a consent decree, the DOJ declares your department is being run outside the normal parameters of a police department, so the FBI is brought in to run and supervise your department—every police stop requires a three-page report detailing the race of the suspect, the reason for the stop, whether it was justified, etc.

A few months later, I found out the FBI was now investigating *my* case—the Alvarez case. Curious, I got in touch with an administrator at the Fairplex. "Did the FBI come out there?"

He nodded. "Yeah. Two female agents—looked like Charlie's Angels—came out here taking measurements, collecting all the liquor sales receipts for that entire day."

I stared at him for a few seconds, stunned and confused. And I realized—*this wasn't over.*

CHAPTER TWO

GROWING UP SOCAL

Cinnamon Dentyne and a Seventies Ford Pickup

I met my dad on a ballfield.

I was eight years old, playing second base. At the time, I was living in my grandparents' house with my mom and older brother. To be honest, I never knew I had a dad until that day. I'm not sure how I thought I'd come into existence, but I'd been more concerned about other things as a kid.

After school, I'd take the bus over to the fields to play in the Boys Christian League on Monrovia's south side. On this particular day, after the ninth inning closed with a pop-up fly ball to the center fielder, I saw my older brother, Greg (three years my senior and my only full-blood brother), walking up behind the chain-link backstop with a man I'd never seen before. Greg introduced him: "Hey, this is our dad." Say what? Was I supposed to just believe that this guy was my dad? My brother apparently believed he was. Awkward to say the least. What were we supposed to say to each other? None of us really knew. My dad was a man of few words at the time, and we were just kids, so we fumbled through small talk and shuffled out toward the parking lot. I saw my grandpa's old pickup at the far end (he had loaned it to my dad), with all my worldly possessions in the back—two black trash bags of clothes, a box of miscellaneous stuff, and my red Schwinn Sting-Ray bicycle—along with my brothers'. Oddly, I wasn't all that surprised; we didn't exactly have a Norman Rockwell home life. I tossed my baseball glove next to my bike, and my brother gave me a five-word explanation: "We ain't going home, bud."

I wasn't happy about the situation and was still uncertain about whether this man was really my dad. And now I was supposed to go live with him? It was great living with our grandparents, even if it wasn't typical. But apparently, my future was being written by another hand, so I got in the truck to see where it would take me. We motored over to my grandpa's corn distribution business at the other end of town. My grandparents' families were immigrants from Denmark and had settled into the business of trucking corn from California to companies in other states that made corn chips and tortillas. There, I learned that my dad was an over-the-road truck driver for Valley Grain Products in Madera, up in the San Joaquin Valley. He was driving a classic cabover Kenworth with a big sleeper that I thought was all kinds of cool. Even better, he'd rigged up the rack on the deck behind the cab for our bikes. He stowed our bags inside the truck, and we took off. I never got to say goodbye to my grandparents or friends—we just disappeared.

For six hours, we rolled north on I-5 and Hwy 99 to Madera, volleying sporadic conversation but mostly fidgeting in uncomfortable silence. For distraction, I stared at all the lights and switches and buttons in the cab—I'd never seen anything like it. Dad looked like a salty trucker from the movies, one hand on that huge steering wheel and the other on the shift lever. Even better, we heard the CB radio squawking with secret codes and strange names they called "handles." As we were slogging up the Grapevine in Central California at about nine miles an hour, another big rig passed us, and we heard, "Hey, Caterpillar, you gonna ride a bike when your truck can't make it up this hill?" Now we knew Dad had a cool nickname, too, one we used generously over the years.

Finally, we turned into VGP's dusty driveway and transferred all the gear into Dad's green-and-white early 1970s Ford pickup truck.

From there, it was another half hour to his house in the suburbs across town. I'll never forget the last thirty minutes of that trip. At one point, Dad reached over to open the ashtray. Inside was a little five-pack of cinnamon Dentyne gum, his favorite. He pulled it out, and we all took a piece, and a few miles later, the inside of the truck smelled like cinnamon. It was one of those poignant moments that sticks with you. To this day, the smell of cinnamon takes me back to the day I got to know my father.

Things turned somewhat less storybook after we pulled up to a nondescript single-story house on Lansing Street, a place I will never forget. I had never met my half-brothers, either, and they all filed out of the house at once—three-year-old Grady waddled out in his diaper, followed by Jed (two years my junior) and Shaun (four years my senior). Behind them was my stepmom, Kathy, passing around those courtesy hugs and sentiments devoid of any real meaning or connection: "I can't believe you're here! You look just like your mom. How was the trip? I'm so excited to have you live here with us." She followed her overcooked posturing with a disapproving, very "non-cinnamon-y" side-eye.

Greg and I both had shoulder-length hair at the time, because, well, that was what you did in the seventies, and our mom, God bless her, hadn't cared. She'd had far more weighty problems. We weren't in the house five minutes when Kathy dragged a wooden chair across the kitchen tile and said, "Have a seat." It wasn't a request—protests or opinions were strictly forbidden. So I reluctantly took my perch, shoulders slouched, face engraved with a frown. Turned out Kathy was a hair stylist by trade and went after me with a vengeance. She grabbed my head, said, "Be still," and doused me with a spray bottle of ice-cold water. Then the scissors and electric trimmer went to work until I was left with a standard "little white boy" haircut. I felt like Samson—defeated. My brother got the same treatment. We looked like we were ready for basic training and lurched out of the kitchen, thinking the worst was over.

It wasn't. My bags of clothes and box of priceless (to me) stuff sat on the back porch, within sight of the haircutting carnage. "Let's go through this," Kathy announced. Inside that box was my extensive collection of two KISS albums. I loved that band back in the day, but Kathy did not, what with all the satanic undertones and certain condemnation to hell. With great ceremony, she said, "We will not have this in this house," and, right in front of me, snapped the albums in half and threw them in the trash! I couldn't believe what was happening and dreaded what might come next. Riffling through my clothes bags, she found my treasured KISS T-shirt, black with an old KISS Army iron-on. I'd worn that thing the entire summer and never washed it in fear of the picture coming off. I still have a photo of me wearing it at a family reunion at my grandpa's house. None of that mattered to Kathy; she cut it in half, right up the middle, and chucked it in with the shattered albums.

Within an hour of being in a new home, the most beloved elements of my life had become part of the weekly garbage pile (or were drifting around on the kitchen floor). It was a brutal welcome and the start of seven years of what felt like hell on earth. But we escaped when we could, hitting out on our bikes to adventure wherever our wheels would take us. One of our favorites was ripping out to the Fresno riverbed with bags of tortilla chips from Valley Grain in our backpacks. We'd blast off jumps, skid in the dirt, and eat chips for lunch. Gone until dark. It was a great time.

A little analog clock radio helped me get through it too, with just enough reception to pipe in Y94 on the FM dial. Music like Cheap Trick, Supertramp, Billy Joel, Kenny Rogers, and Juice Newton got me through the shitty parental treatment and absence of any friends. Most of the time, I would get in trouble for some ridiculous reason and be told, "Go sit on your bed," and I would spend the entire day quartered to my room listening to Casey Kasem's *American Top 40* all day with that puny radio on full volume. When Kasem's show ended and the gloom returned, I always reminded myself that the sun would rise tomorrow.

Kathy, of course, thought she could save me, and the Church of Jesus Christ of Latter-day Saints was just the thing. A hardcore Mormon attempting to escape her own troubled childhood and atone for who knows what semblance of transgressions, she herded my brother and me into three weeks of missionary lessons, baptism ceremonies, and various other indoctrinations into the faith.

Whoa, rein in the wagons a bit, ma'am—I hear another call. I had a feral streak early on, altogether LDS-divergent, but, not wanting to garner unwanted attention from my step-monster, I was perfectly content to follow its whims. Like the day of the pizza-eating contest at Madera's Fourth of July celebration. I was fourteen, five years into dutifully following all the church rules, including a twenty-four-hour fast on the first Sunday of every month, which happened to coincide this particular year with the town festivities. I liked having fun, chasing girls, and other rambunctious teenager stuff, and Jed was right there with me. We were great friends, and by this time, we had a little more freedom from home internment. What better destination than a town carnival, so after church, we headed straight to the big event. Somewhere between the Tilt-a-Whirl and corn dog stand, we spotted a sign announcing something we couldn't resist:

> **Pizza-Eating Contest 3 P.M.**
> **WIN PRIZES!**
> **Sign up here!**

I was starving, and the thought of scarfing down piles of pizza sounded amazing. Jed said, "Yeah, do it!" I signed up, and at 3 p.m. sharp, I sat at a long table with a red, white, and blue tablecloth and five would-be competitors. Five minutes later, a steaming pepperoni pie from Straw Hat Pizza appeared in front of me, destined for rapid consumption. I ate the whole thing, chased with intermittent guzzles of Pepsi (another LDS violation), and walked off with a trophy emblazoned with, "Boy, did I eat a lot of pizza! Madera Fourth of July celebration, 1981."

Walking home, I had eaters' remorse. "I don't know if that was a good idea."

Jed reassured me, "Dude, it's cool. Just hide the trophy."

But when we got home, we let it slip, and Kathy went berserk. "You let down the Lord. You owe Him an apology. You're not a good person." I was, of course, grounded for a month—no going anywhere except riding my bike to work at my busboy job at the Vineyard Restaurant in town. I had a shift that evening, a few hours after the pizza contest, and got sick halfway through. The boss didn't like that I threw up on some of the dishes and fired me on the spot.

Kathy didn't like it either. "Congratulations, Chad. You've proven to me that you can't be trusted."

The next day, she handed me a spoon and a fork, which seemed rather strange. "Here, you're going to go weed that garden, and you're not doing anything else until it's done."

I stared at the utensils in my hand. I walked slowly out to the garden, a traditionally huge Mormon rendition, about twenty by fifty feet, that supplied all our vegetables. To me, it looked like an NFL football field choked with weeds, and there I was, with a fork and spoon. Not a hoe or shovel—a freaking fork and spoon! I toiled out there for a solid week and, of course, never came close to getting done. The fork didn't survive, but I'm pretty sure Kathy put the spoon back in the drawer.

For the first few years I was at my dad's, my mom would visit my brother and me in Madera three or four times a year. She lived three hours away, and my stepmom never made it easy. I would always get a ration of crap whenever I came home from a visit with my mom. We'd meet up at a park or somewhere in town, but that came to a screeching halt over a new pair of jeans.

A quick look at a day in the life of Chad: I'd get up early at five every morning and put on my one pair of hand-me-down corduroy jeans with holes in them and my shoes with more holes, have scripture study, then make breakfast for everyone (old-fashioned boiled oatmeal). Afterward, I'd clean the kitchen counters and sink with bleach, then go to school reeking of the stuff all day. I remember a kid named Michael Wong always made fun of me and called me Ajax, and of course, other kids did too, but that's how kids are . . . just plain ruthless. I recall when, during a surprise visit from my mom when I was about twelve, my mom picked me up after school, saw my state of garment disrepair, and took me straight to JCPenney. I walked out in a new pair of Toughskins, the kind with patches on the knees, and knock-off sneakers. Hardly GQ dapper, but the most inwardly confident I'd felt in a long time.

Until I got home. Kathy took one look and said, "Oh, it appears your fairy godmother came today. Take off those pants and shoes right now." She called for Jed and gave everything to him. "These are now Jed's, not yours. Now he can see how it feels to have a fairy godmother for a change." I could tell he felt horrible, so I just shuffled down the hall in my underwear and lost myself in little radio music. What else could I do?

Later that evening, I heard my dad and Kathy arguing, then Dad's footsteps to my room. He stood in the doorway with a simple proclamation: "Your mom's dead to you. You're never seeing her again, so get over it. Don't see her, or there'll be trouble."

My mom called bullshit on that and snuck visits at the ballfield or some hidden place after school. Unbeknownst to me, she had been sending letters with five-dollar bills tucked inside for three years. I never got a single one. Turned out, as I was told, my oldest stepbrother, Shaun, troubled in his own right, had swiped the letters from the mailbox and pocketed the cash. Kathy found out and went through the roof, raving mad, calling us deceptive little devils, and on and on.

Walking home from school one day during my sophomore year of high school, Kathy pulled up in her fire-engine red '77 Ford Thunderbird.

"Get in the car," she yelled. I got in. That woman scared me.

She hightailed it home, where my dad's truck was in the driveway. What the hell? He was never home early. In the back of the truck were two trash bags of clothes, a box of stuff (minus the KISS albums), and my red Schwinn. She had found out that my mom was secretly seeing me and was pissed. "Get in the truck and stay there," she barked, then got the keys from inside and drove me two and a half hours, in dead silence, to a duplex on Junipero Street in Monterey. Kathy slammed the truck to a halt at the curb, climbed into the back, threw my bags and bike on the ground, pulled me out of the passenger side, and took off—no words. Once again, no goodbyes to anyone at home, school, or church. I never got to say goodbye to any of my stepbrothers, my dad, or Greg. I just vanished.

I sat there on the curb alone for an hour, maybe two, staring at the pavement, thinking this life was nothing like the classic California surfing, hot rods, and beaches, when my mom's husband at the time pulled in. I had met Jon one time before, years ago, out in Pasadena, so at least he knew who I was. We looked at each other in shared bewilderment, and then he walked over and asked, "What are you doing here?"

"I don't know what I'm doing here." We parked my bike around back and brought my bags into the house, and before long, my mom got home. It turned out Kathy had called her and, in effect, said, "You can have him. He's nothing. He'll never be anything." But my mom never gave up on me. I stayed there with her a while until Jon and my mom divorced. Soon thereafter, my mom got remarried to a man named Mike, and sadly, we didn't get along. Ultimately, my mom sent me to Riverside to live with my uncle and cousins during my senior year of high school. Then, well, you know the story: It was off to Utah with my high school girlfriend and having a baby there, then moving back to Riverside and landing a job at Food 4 Less. All this while having another baby and honing a knack for making one relationship mistake after another.

Now, a year later and in my second marriage, I was promoted to store manager at Food 4 Less in Pomona. It was 1993, shortly after the Rodney King riots. This particular store was located just outside an especially rough part of town known for gang activity and related disorder. One resident thief strolled into the store almost daily to steal a bottle of gin. I had finally had it with this guy, and the next time he swiped a bottle, I ran out and tackled him in the parking lot. There we were, wrestling in front of the store as patrons gathered around to watch. I now look back and see this as a premonition of the county fair incident in my future. I was wearing my official Food 4 Less uniform—a white polo and slacks—with a big ring of store keys clattering at my waist, gradually losing the match (technically, he was handily beating my ass), when an unexpected Good Samaritan saved the day. We had a Cubano guy who just hung around the front of our store. He never interacted with customers or caused any trouble. In fact, after a while, he sort of blended in with the scenery—he could've been a store greeter or rung a Salvation Army bell at Christmas. I never made him leave or hassled him, and I guess he appreciated that because he hustled over and broke up the fight enough for the thief to stop pummeling me and hightail it out of there. I picked up the broken bottle pieces off the ground, looked at my torn shirt and pants, and realized how lucky I was that our resident "referee" had saved my ass from serious damage.

At the time, I was in the Riverside County Sheriff's reserve training program, and when the Pomona police officer showed up at the store a few minutes later, one of them (I ended up working with him years later) asked what the hell I'd been thinking, risking my life over a bottle of lowbrow gin. I didn't have a good answer—I obviously wasn't thinking, but I couldn't stand thieves and just had to put a stop to that gin-stealer. But this Food 4 Less WrestleMania referee relationship between myself and Pomona PD didn't end there.

A dozen years later, working graveyards as a Pomona police officer, my radio broadcast that a crime of some type was going down around Garfield Park. I rolled over and noticed a Hispanic male sitting on the curb in front of a used car dealership adjacent to the park. Maybe he'd seen something or been involved somehow. He didn't speak English, and shortly into my questioning I said, "Wait a second, I know you." I called over one of my Spanish-speaking officers to ask the guy if he remembered me or knew who I was, and sure enough, he did. He was my storefront savior—of all the people to run into in the middle of the night on Holt Boulevard in Pomona. My partner translated my sentiments: "I never got a chance to thank you. If you ever get in trouble, call me, I owe you one."

I never saw him again after that day, but it was one of those powerful, circle-back moments. I had come so far since that day in the parking lot. Seeing him sitting there took me back, not only to that day but to all the days that led me here. That was me in the cab of my dad's truck, searching for something to say. That was me in Kathy's car, not knowing what would happen next. That was me in front of my mom's house, kicked to the curb.

But I'd known I couldn't stay on that curb forever, so I'd gotten up and decided to live my dream.

CHAPTER THREE

LIVING MY DREAM

CHiPs and a Brain on a Bed

I started at the Rio Hondo Police Academy in Class 140 in July of 1997 as one of three recruits for the Pomona Police Department, along with Bill and Christine. Located in Whittier, southeast of downtown Los Angeles, this academy location earned its nickname and reputation, "the death academy," from the stifling hot summer temperatures when instructors sent recruits through daily conditioning and fitness tests like agility drills, obstacle courses, and endurance running. Effective, yes, but if I never have to run around with full gear and a Kevlar vest again, it'll be too soon. Even so, I actually loved the academy experience. The physical fitness was challenging but not impossible. I was lucky to have my academy partners there, as we were always motivating each other and cracking jokes. Every Friday on the two-hour drive home, Bill and I would stop at El Boca del Rio on Beverly Boulevard in Pico Rivera to get the best Mexican food in the world! Having gone through Riverside's reserve training a couple of years prior, I excelled at the academic portion, but a recruit from Pasadena beat me out of the top recruit spot; he was a stud. The tactical staff was particularly tough, and I think they enjoyed seeing us suffer, but they prepared us well for the job we were about to take on. The academy definitely lived up to the banner across the entrance: "Only the strong survive."

But we made it. On December 4, I graduated with my academy partners, Bill and Christine, and we were sworn in as full-time officers by Chief Shaurette with the Pomona PD. After the ceremony, the chief sent us home for the weekend with our first marching orders: "Show up at the training center on Monday morning at eight o'clock. Don't be late. Report to Sergeant Johnson."

I enjoyed a relaxing weekend with my wife, Kimberly, and when Monday morning arrived, I was out the door with a spring in my step and an easy thirty-minute drive to my first official day as a police officer. I figured I could leave by 7:15 and still have plenty of time to get there by eight. I hopped in my white '97 Dodge pickup, got on the 60 freeway, and cruised westbound toward Reservoir Street, but Southern California traffic had other ideas, backing up in a hurry. It was going to be a mess. I glanced at my watch with the exit ramp in sight—five minutes to get to the training center. *Shit, I'm not going to make it.*

The two-lane off-ramp was choked with vehicular expiry—stone-faced drivers in hundreds of steel tombs frozen in place. Nothing was moving—and I needed to, right now. Maybe I could dodge it with some deft maneuvering, I thought. I blasted down the left turn lane, forced my way around the stopped traffic, and nearly ran into a Pomona PD motorcycle cop standing in the middle of the road next to his big Kawasaki. He was all business, menacing—dark sunglasses, helmet, huge caterpillar mustache. He stared me down, blew his whistle, and pointed me to the curb. I obliged, parked, and waited with my hands at ten and two as the officer pulled in behind me, his bike's red and blue lights flashing. All I could think of was that I was about to get fired on my first day of work as I watched this guy in my side mirror, walking to my door. (Objects in mirror are closer than they appear.) He sidled up, put his hands on my door, and asked, "You know why I stopped you?"

"Yes, sir, I have no excuse. I'm late to meet Sergeant Johnson at the training center."

He raised an eyebrow. "Do you have your ID with you?" I handed him my driver's license and Pomona PD identification.

"Jensen, huh? I've seen your name before. You're third from the bottom. I'm Leonard, three from the top." He walked back to his bike, came back a few minutes later, and handed my ID cards back. "Tell Sergeant Johnson that you and I just had a conversation."

"Yes, sir, I will." And with that, I hurried (safely) on, rolling into the training center ten minutes late. *I'm screwed.*

I walked into Sergeant Johnson's office, greeted by the smell of stale bourbon and cigarette smoke. Johnson was a quintessential, crusty old beat cop who'd seen it all from behind the badge since the sixties. "Sir, can I talk to you for a second?" I asked.

"Sure," he grunted, puffing on a cigarette.

I explained the whole story in my most despondent "please don't fire me" testimony. "Corporal Leonard stopped me and said to tell you we had a conversation. I'm sorry, sir. I've wanted to be a police officer for so long, I got through the academy, and now this. Am I going to get fired?"

"Well, did you learn anything?" he asked.

"Yes, sir, I learned I need to get off at a different exit."

"Good," he said. "Go sit with your partners."

Bill and Christine waited for me in the training room, a thirty-by-thirty-foot space with a set of double doors open to the west. It was a beautiful morning, and we sat there like three young kids who'd been sent to their room. Bill, a stoic former marine, finally broke the silence: "Damn, Jensen, you really fucked up."

No one said much else for twenty minutes or so. Birds chirped outside in the early morning sunshine, and a crystal-blue sky heralded a stunning winter day ahead. All in all, quite peaceful—until I heard the unmistakable roar of another Kawasaki police motorcycle out there somewhere. Getting closer.

I tensed up. Something wicked this way comes. The motorcycle pulled up right outside the open doors, piloted by a big, burly monster of a man. Sergeant Gillespie killed the engine, levered the kickstand down, and climbed off as the bike settled into repose amid an orchestra of creaks, pops, and groans. I can still hear the muffled rub and twist of Gillespie's leather boots and gear belt as he walked around to the right-side saddlebag. He flipped the buckles open, reached in, and pulled out the latest unabridged California Vehicle Code—a three-inch-thick heavyweight ideal for a relaxing evening read. He walked up the steps and through the open doors with leather from his tall motor boots and duty belt creaking and popping. At some point during all this, I said to Bill and Christine under my breath, "I'm really sorry for whatever's about to happen."

Gillespie seemed to grow into even more of a giant as he got closer, like a frigging horror movie. Then he just stopped and stood there in the doorway (for dramatic pause if nothing else), blocking out the sun, staring at us. Finally, he barked in his deep baritone, "Which one of you is Jennings?"

I hesitated for a second—I mean, my name isn't Jennings—but then realized he was obviously talking to me. "Yes, sir, I think you mean me."

Without another word, he launched into a scathing tirade that sounded like one of Bill's marine drill instructors. "You, young man, owe me a five-page paper on why police officers need to obey the vehicle code. Have it to Sergeant Johnson tomorrow at 0900!" Bill and Christine just looked on in silence, collateral damage to my verbal lashing. Gillespie slammed the vehicle code on the desk in front of me—it boomed like Dirty Harry's .44 Magnum—as he finished his "motivational speech" and walked out, stalking down the steps and onto that big Kawasaki. I could swear I was looking at Arnold Schwarzenegger in *The Terminator* as he fired up the bike and rode off.

I was in shock. I couldn't believe how badly this had escalated. The whole training center would hear about this, most likely the entire department, and I'd probably never live it down. But then I heard a distinct laugh outside that I recognized as Corporal Edwin's, the volunteer and training officer coordinator at the time. What was so funny? Did I have a chance to redeem myself? I had plenty of time to stew; they left us to sit there another couple of hours, without a word, until it was time for lunch and a few mundane tasks that soon closed the day. I went home to start my homework.

The next morning, I turned in the five-page paper to Sergeant Johnson. "Sir, here's the paper that Sergeant Gillespie wanted."

Sergeant Johnson took one look at it, threw it in the trash, and asked again, "Did you learn anything?"

"Yes, sir," I replied, relieved. "I just need to leave earlier."

"Good," he said. "You're fine. Relax."

About a month later, I ran into Corporal Leonard at the station. We were dressing to go 10-8 (available for duty), and he walked over to my locker. "Hey, I just want to let you know, I didn't want anything like that to happen. I didn't think Gillespie would do that," he said. I appreciated the gesture and eventually worked with Corporal Leonard later on. He turned out to be a really good guy and one of my favorite mentors.

Another one of the sergeants I worked for, Sergeant Jones, was a brilliant mind. He could solve quadratic equations and tell you the chemical breakdowns of stuff, but he often seemed forgetful with names. He thought my name was Todd and used it all the time, so I just went with it. Around the station, and even at home for a while, I was "Todd Jennings." (Amiee still calls me Jennings, especially when I'm in trouble.)

About a year or two after Training Day One, I was out on my own after completing field training. Gillespie, now a lieutenant, was working as the watch commander at the station one day, and I was at records turning in a report. I saw him and tried my best to avoid eye contact, hiding around corners and pretending to drop things as

I worked my way out of the building. Just when I thought I was in the clear, he spotted me and called me into the office. He thundered a command down the hall at me. "Jensen, get in here!" *Shit. I'm getting smoked again.*

I stood tall in his office, my knees shaking. "I heard you do a hell of an impression of me, and I'd like to see it," he said with a smirk.

I played it cool and respectful. "Sir, I'd love to do it at your retirement party, but I wouldn't feel comfortable doing it right now."

"I'll remember you said that," he replied, before dismissing me with a big smile.

Fifteen years down the road, he was now Captain Gillespie, and his secretary, Janet, summoned me to his office. It was like getting called to the principal's office in grade school. I thought I was in trouble again, sweating bullets on the way to her desk, but she met me with a big grin. "I have a favor to ask."

"You name it," I said.

"You have to do that impression of Gillespie at his retirement party next week."

After the initial dread wore off a few days later, I agreed to do it. Janet even had the party's program printed with "Special Presentation by Chad Jensen." I had to do this right. I borrowed a beach cruiser from one of the venue's dishwashers, taped some DIY sergeant's stripes to my sleeves, donned a fake moustache and a motor helmet, and rode down the hallway right into the ballroom. Two hundred people went dead silent. I parked the bike and did the whole Gillespie routine, right down to the vehicle code manual and "Which one of you is Jennings?!"

The room erupted in laughter, and I topped off my presentation by making motorcycle noises while pedaling around the tables and "leaving the scene" back down the hallway. It was great fun, and later that evening, standing at a men's room urinal, guess who walks up next to me? None other than Gillespie. He looked over and said, "That was pretty good."

Along with the revelry came a turning point for me. I had found my place by being who I am and staying true to myself, no matter the circumstances. This became the life mantra that gave me the strength and wherewithal I needed to make it through debilitating challenges to where I am today.

Left or Right on Mission

Moments like the beach cruiser retirement party were, of course, not the norm. In fact, a police officer's day-to-day is made up of much more sobering incidents. One of my first field training officers (FTOs) was Corporal Robert Devee, a larger-than-life guy like Gillespie. I spent six weeks training with him, and his powerful, trademark message really hit home with me: "Every day that we go 10-8, our number one goal is to come home safe. Things can go wrong on this job, and any day could be your last day. It could be as simple as if you turn left or right on Mission." The times and experiences I spent with Devee are among some of my most special memories in my career. I absolutely appreciated his method of teaching me during my training years and his constant counseling and guidance throughout my career. One of the best things he ever said to me was, "Chad, if I'm the voice of reason, you should probably listen."

Pomona PD's squad car parking lot exited onto Mission Boulevard, so when you started your shift, you had to turn left or right onto Mission. Simply put, your life could be changed based on which way you turn that day. Years later, as I held tenure as a field training officer, training dozens of officers over the years, I passed down that same story to them.

A gut-wrenching example of timing and decisions unfolded on the Pomona courthouse steps in broad daylight. Santa Fe Springs area California Highway Patrol (CHP) officers, along with other CHP and allied police agencies, conducted traffic court testimonies on Wednesdays in Pomona. Officer Thomas Steiner, working commercial vehicle enforcement that day and dressed in the standard blue jumpsuit, was walking out the front doors of the courthouse (next door to our police department) around 2:30 p.m. when a sixteen-year-old wannabe gangster pulled up, pointed a revolver out the driver's window, and shot at him five times. Steiner was not initially struck by any bullets as he unknowingly walked toward the suspect's vehicle, but when he turned away from the suspect, one bullet struck him in the back of the head, killing him instantly.

I was working that day around the corner as a motor cop, not even two blocks away, conducting a traffic stop and wearing the exact same type of uniform as Steiner since I also did commercial vehicle enforcement for our department at the time. A call went out reporting an officer down on the south side of the courthouse, based solely on 911 calls since no one was on our radio frequency. I jumped on my bike and hauled ass over there, the second person to arrive at the scene. It was my first time seeing a fellow officer bloodied and hurt and helpless. My heart sank and broke, the shock immediate and sickening. This was something that I never thought I would see: a fellow officer lying in a pool of blood. He was lying there being held by the courtroom bailiff, who'd heard the shots. Soon after, LA County Fire paramedics arrived and started performing chest compressions, but they couldn't save him. Everyone began to step back, with nothing more to be done. That mental snapshot of him, lying still while surrounded by grave faces, remains as vivid as the day it happened.

After an exhaustive sixteen-hour search with hundreds of mutual aid officers from other agencies assisting, we eventually found and arrested the assailant just a few blocks away. When questioned why he would shoot an officer who was simply walking out of a building, he said he was trying to "earn his bones" to become a 12th Street gangster. He believed killing a cop would make him a big shot. When asked why he chose Steiner specifically, he admitted he had driven around for a half-hour looking for any cop to shoot. He didn't realize Steiner was CHP—he thought he was a Pomona officer because we wore similar blue uniforms.

This realization hit me hard. He could have easily rolled up and targeted me instead. It was pure coincidence that I hadn't been there. If I had been on those courthouse steps instead of two blocks away, I would be dead.

SWAT

No-knock search warrants, breaching doors, flash bangs—every day as a cop is risky enough, but SWAT adds exponentially more danger on top of it. One particularly tense experience had me reconsidering whether it was all worth it.

We were serving a no-knock warrant for guns at a gang crash pad house. It was three in the morning, and we were going in hot. The plan was for a double entry: hit the front and breach the back at the same time. The back limited breach was supposed to be a distraction—smash some windows and make noise to get the bad guys' attention so their guns wouldn't be aimed at us when we breached the front door. We were all on comms so we could talk during the raid.

It started out simple: I ran up to the front door with my team, and we executed a textbook breach. *Boom! Boom!* Two strikes—the door blew open, we dropped a flash bang, and we were in, saying hello to a small group of gangsters lying around on couches and other ratty furniture. We took them into custody without incident, seized the guns, and started back to the office for debriefing. But something didn't sit right with me—I hadn't heard any flash bangs or glass breaking from the back breach before I went through the front. I didn't remember hearing anything at all. I checked with the rear team, and they said, "Yeah, we decided not to do that." What the hell? I was standing in front of an open door with a sixty-pound battering ram, completely exposed, no gun in hand, and they hadn't even followed through on the plan that was supposed to keep us safe.

Over the years, I had kept my mouth shut on things I knew were wrong because I was the new guy or not the team leader, but that day, it was different. I told them flat out how ridiculous this was—we didn't communicate, we had outdated equipment, and we didn't train nearly enough. As a part-time team, we didn't devote enough time to perfecting our craft like the top-level teams, LAPD Metro and LA County Sheriff's SEB, did. With the mistakes that were being made, it felt like someone was going to get seriously hurt or killed. Tragically, a short time later, my friend and former teammate Shaun Diamond was killed in a breach operation similar to mine—the exact thing I had warned them about. According to some of the team members, the breacher was not notified of movement inside the house prior to breaching. My breaching experience dictates that interior movement should be a call for "compromise." If the element of surprise is gone, no breach should occur. I do not want to comment further, because I wasn't there at that door. However, the limited debrief information that was provided painted a picture of a lack of communication and operators being put in harm's way unnecessarily. The truly tragic part was that the department never debriefed its personnel on what had happened. It was the same old story: Shit happened, and no one talked about it or learned from it. Just swept it under the rug again. From Danny Fraembs to Shaun Diamond to Greg Casillas, the department never provided a clear picture of what had happened. Even the top-tier SWAT teams like LAPD Metro and LA County SEB have in-depth debriefs where they show exactly what occurred and what can be learned from it. Isn't that part of honoring those who have paid the ultimate sacrifice? To learn from our mistakes?

Jack-in-the-Box Girl

Another breach operation, that fortunately did not result in a fatality, occurred during an early morning search warrant. These warrant events started around two in the morning when the team met at the SWAT office to gear up and brief for a four o'clock service. Everyone was usually glad for the work and stoked to get in the field, dressing out and listening to some hype music like we were going to the state football championship. Guys passed around Skoal tins, Monster Energy drinks, and Red Bulls while the old-timers dipped Copenhagen and chugged black coffee. After the brief, every member of the team slapped their hand on the team placard above the team office exit door, a fifteen-by-fifteen-inch wooden plaque we had made that displayed an Edmund Burke quote: "The only thing required for the triumph of evil is for good men to do nothing."

We had a preliminary buddy check for critical equipment readiness, walked through the entry stack, then loaded up, this time for a ten-minute ride to the pre-stage a couple of blocks from the target. Once inside the armored vehicle (ARV), nerves were on full, fast-twitch alert, but the second we arrived, the crew hopped out, and everyone knew their job, from outer perimeter and breaching to the arrest team and stack guys (the entry team). It was a well-orchestrated event that usually ended in success, with no one hurt and the bad guys in custody. But as the saying goes in SWAT, "Sometimes you get the bull, and sometimes the bull gets you."

This particular homicide search warrant was for a southside gang member. Our team scout had confirmed the target was at the house, so as usual with these types, there was a high probability of a gunfight—a real shit-sandwich—so we had to be prepared and alert. M-4 at high ready, I was now the number one in the stack, in the lead as guys peeled off and worked through the house to the last bedroom at the end of the hall. No contact with anyone in the house yet; the bad guy had to be in that last room. I booted the flimsy, hollow door, which opened into a ten-by-ten-foot room with a queen bed in the middle, directly in front of us. Typical of the neighborhood, there was garbage and other debris all over the floor and no sheets or covers on the bed—just a nasty, putrid old mattress. Oddly enough, with all manner of more pressing thoughts of tactics and staying alive, I wondered, *Does someone actually sleep on that thing?*

We swept to clear the hard corners, and as we turned back to clear the far side of the bed, a teenage girl, roughly age twelve to fourteen, popped up from the ground on the far side of the bed and like a jack (jill)-in-the-box. With her long hair flying and her hands up, she yelled, "I'm here!"

Holy shit! She scared the hell out of us. Thanks to many years of experience and training, as well as good trigger discipline, no one fired off a round. Ultimately, the suspect was found hiding in the attic, without weapons (likely because he hadn't had time to grab one). There were no shots fired and no injuries to anyone; we just aged ten years after the scare. Later, I spoke to the girl on the curb outside and asked her to yell out next time that she was in the house, really loud, *before* she jumped up and gave the police a heart attack. Although we got the shit scared out of us by a pre-teen, at the end of that day, we got the bull.

Check the Doorknob, Dummy

We had orders to hit a dope house southeast of the city one night, and we arrived to see a set of large French doors, which are generally very easy to breach. We dialed in a good tactical approach with barely a sound and called up for the breacher (me). I was psyched while walking up there with our thirty-pound breaching tool—me big strong man, break door, save day. But I hit the door, and nothing happened. Then I hit it five more times, and the thing still didn't open! I was getting pissed when one of the team vets stepped up and said, "Try the doorknob, dummy." I reached out to turn it, and sure enough, the door opened like a butler was waiting to usher us in.

The lesson that night was to slow down, think more, and act a little less. Getting all worked up for a fight usually inspires tunnel vision and can go on to result in a failed mission or worse, someone being hurt or killed.

That experience, and so many like it, drives home the depth of camaraderie in SWAT circles. You won't find a better group of guys to go to war with—but if you screwed up, we were *merciless*. The ribbing was harsh and came fast, but it always came from a place of love and trust.

The Greek King Leonidas quote tells our story well: "Lessons not learned in blood are easily forgotten." For us, this meant blood in the proverbial sense, through embarrassment and ridicule. And honestly, I wouldn't change it for anything. You never knew when your turn was coming; it was like a room full of rabid pit bulls, just waiting to turn on each other. But here's the thing: If the team didn't trust you, you *wouldn't* get the ribbing. They'd just work around you, and that silence said everything.

That's the part you miss most when you retire.

Ponch and Jon

I don't think I ever missed a single episode of *CHiPs* back in the day. Even just the show intro had a cool vibe, the way the riders shifted gears on the big Kawasaki bikes with the backs of their boots on the heel shifter. I always wanted to be one of those guys, and as a new officer on the streets, around 1999, I focused a lot on traffic because my ultimate goal was to become a motor cop. At the time, most of my traffic stops were out in the Beat 1 section of town, an area patrolled by one of the veteran cops. Sometimes, we'd overlap shifts, and I'd end up doing traffic in his beat, which really pissed him off. He was out there dealing with gangsters, guns, and dope busts, and here came this knucklehead new cop pulling over cars all the time, scaring the fish away.

And I had no idea. I was just out there having a blast, feeling like Batman, crushing traffic crime. Like a typical new guy, I thought this was the greatest job ever, getting paid to pull people over and write tickets. Finally, this hardened vet had had enough. He pilfered some copier paper from the office, made up little "campaign posters"—*Jensen for Motors*—and posted them all around the station in a lighthearted push for me to go to motors. There were a couple of openings for motor officers, and I'd already put in for it, but you typically needed at least five years on patrol to be considered. Somehow, maybe because of my excessive ticket writing and traffic enforcement, I got selected, and at the same time, Bert, who had about three years on me, got selected too. We both ended up being sent to motor school at the Los Angeles County Sheriff's Department's Emergency Vehicle Operations Center (EVOC) program at the (post-dated irony) LA County Fairgrounds. I'd been riding since age five, mostly off-road, and couldn't wait to get started.

We had a blast. Bert's infectious sense of humor had us cracking up every five minutes. Prior to the start of motor school, our corporal took us out for some practice runs around the city. Bert had never ridden a motorcycle before and had to borrow a silly little enduro from one of his buddies to get his M1 license at the DMV. Once we got into real training with the big, heavy Kawasaki 1000s, especially the cone patterns, it wasn't easy for him. We spent hours weaving through specially set-up orange traffic cones in serpentine patterns, doing very difficult, sharp U-turns—and falling repeatedly.

At one point, the corporal announced that we needed to be able to chase bad guys off-road in the park areas of the city, leading us to a dirt path up a steep hill. I went first and flew right up behind him to the top, where we parked to cheer on Bert. "C'mon, man, go for it! You can do it!" Hardly a quarter of the way up, he lost control, tumbled a bit, dumped the bike, and rolled back down the hill. But true to Bert form, he made it fun. He got up, slapped dirt and twigs off his uniform, and shouted up to us, "Can't we just go get a lemonade already?"

Eventually, we both graduated motor school and hit the road as motor cops—the *CHiPs* dream come true for me. We couldn't have scripted it any better. Bert's Hispanic, and I'm white, so the Ponch and John references kicked in straight away. The motors team consisted of an established older crew and then us, so we spent a lot of time together by design, adding fodder to the TV show duo comparison. In fact, our traffic secretary always used those names instead of our real ones and hung a Ponch and John photo on the traffic office wall, of course with "Chad" and "Bert" written below.

And good ole Corporal Leonard reappeared on the scene. My first day of motor training had me paired with him, running radar out on Valley Boulevard. He rolled back after one of his traffic stops and said, "Alright, you're on the next one, bud." We didn't have to wait long for a car to come speeding down the road. Nothing crazy—maybe 59 in a 45 zone—but I went after it and made the stop, and at the wheel was a quiet, polite, friendly teacher from a local elementary school. I thought, *Oh, no way, I can't write this nice lady a ticket.* "Please slow down, ma'am. Have a nice day."

I motored back to Leonard and told him, "I just couldn't do it—she was a teacher, the nicest lady ever. It would've felt like I was writing my own mom a ticket! So I said, 'Ma'am, just slow down.'"

Leonard gave me a to-the-point field lesson: "You need to get rid of that heart and start dumping ice water in your veins, or you're not going to make it in this job."

I appreciated his guidance from years on the road and put it into play in my own way later, but on that day, it was great to do something nice for someone. It feels good when you're Batman, when you're able to right some wrongs, because nobody's ever happy to see you as a motor cop. Every involvement you have with the public is generally negative in nature. Even then, I always made one thing very clear: Every day I worked, I would do one random act of kindness. And I have my mom to thank for that. She was a big "pay it forward" person and encouraged me to let one person go every day I was out there, on the condition that they did a similar act of kindness for someone else and paid it forward.

I maintained a consummate professional presence my entire career, but there's a place for engaging connection with others as well. It goes a long way, especially when wearing a badge.

Rialto

Not all my experiences in motors were so friendly. One night when I was off duty was very nearly a deadly one. I had divorced Kim before I'd been selected for motors, and by this time, a couple of years later, I had moved in with Amiee, whom I had met on the job in Pomona. She'd been an EMT on an ambulance working the southside, and I'd been working a southside patrol beat as well. We'd dated for a while, ultimately getting engaged and moving in together in Rialto, where my youngest son, Ryder, was born. Rialto was a rough-and-tumble city in the throes of redevelopment at the time. We had bought a small, three-bed, two-bath, 986-square-foot house through a program called "Officer Next Door," in which police officers, teachers, and firefighters could buy a home in a redevelopment area for half price as long as they committed to staying in the house for at least three years. After Amiee and I got married, I spent a lot of time racing motocross, specifically at the Fire and Police Motocross Nationals, which pit firefighters and police officers against each other to entertain lively crowds lining the track and hosted great post-race parties.

Well, one day during practice, I hit a false neutral going up the face of a double jump, planted into the front of the landing, and got myself a Lisfranc fracture, historically common among cavalrymen of the Napoleonic Wars who fell from their horses and got their feet caught in the stirrups, breaking bones at the top of the foot. At my version of a Napoleon battle, I caught my foot between the handlebar and gas tank and ended up in foot surgery, where they covered the broken bones with cadaver tissue, inserted one screw here and another over there, and rounded off the collection with an external rod that traversed the length of my right shin bone and immobilized my foot and ankle to aid in healing.

While recovering at home on a hot summer night a couple of weeks later, with a bunch of metal lodged in me, something told me to pull my SIG Sauer 9mm from its safe. The safe only needed one spin to open, but for some reason, I just felt the gun should be next to me on the nightstand. My wife, who was six months pregnant at the time, was asleep next to me, and our son, Ryder, who was eighteen months at the time, was in his crib in the next room. Around midnight, a loud banging woke us all up; it sounded like a freight train coming from near the front door. Still on the groggy side, I hobbled down the hallway in my gym shorts, gun in hand, as my wife ran in to grab Ryder from his room.

I reached the front door, and someone was outside, yelling and pummeling the "ghetto screen," a reinforced steel screen door common in neighborhoods like that. I could hear the source of the screen door abuse: a Hispanic male, bellowing in Spanish. I don't speak Spanish and didn't understand a word, but I clearly identified myself to get his attention. "I'm an off-duty cop, and I have a gun. Get the fuck outta here! Get away from my house!" The guy didn't stop—just kept screaming, kicking and pulling at the door, trying to get in.

Meanwhile, my wife called 911, and I opened the interior door to put a stop to the crazy chaos. Suddenly, he stopped kicking at the screen door, looked to his right down the front yard, and backed up about ten feet. The house had typical 1950s construction, with vertical decorative windows right next to the front door, covered with closed wooden blinds. A couple of seconds later, this nut took three big steps and jumped through one of those windows! I can only recall the window smashing and a Los Angeles Dodgers baseball cap and red sweatshirt coming through the window, and then I don't remember much after that.

The next thing I knew, my gun was in my hand, and my wife was yelling at me to put it down because the police were outside, but I was already in defense mode. I didn't hear anything else after that. No gunshots in my head, no ringing in my ears. No recollection of sight picture or squeezing the trigger. Nothing. It's amazing how your body goes into preservation mode and the adrenaline takes over, shielding your ears from the sound of gunshots and your mind from the shock of it all.

By this time, Rialto Police had shown up, and I saw them out there wrestling around with the guy in the front yard. He was screaming in pain from gunshot wounds (I had no idea how many times I'd shot him or where) and yelling, "Madre! Madre!" He was still putting up a fight until the officers finally got him onto a gurney and shipped him off to Arrowhead Regional Medical Center.

After that scary, out-of-control chaos, the first police officer to walk inside asked me if I was injured, and I had a sudden dump of emotion and immediately started crying. I couldn't wrap my head around why it was so emotional at the time—maybe it was the genuine fear of not being able to defend my family and the absolute shock of the situation that were overwhelming. I didn't want to shoot the guy, and I explained to the officer how I'd done everything I could to keep this from happening, but he wouldn't stop! The officer talked to me for a few minutes, and you could tell he had been on the streets for a while and was well-versed in dealing with these types of situations. I remember him saying, "Hey, man, my sergeant's going to come over here and talk to you real quick. Just answer the questions she has, but watch what you say. At this point, you have a right to remain silent as well, any time you're involved in an incident like this, so you need to think about that." He further related that people in these types of situations tend to make emotional statements that can be used against them later on down the road.

About ninety minutes later, the San Bernardino County Sheriff's evidence team showed up to examine the scene. When they were finished, the tech brought me over to the broken window, pulled up one of the blind slats, and showed me bullet holes that proved the crazy dude was indeed inside my home. My SIG held fifteen rounds, and there were five missing; I'd hit the guy four times in the upper torso and abdomen, which had caused two collapsed lungs and bowel trauma. He was transported to Arrowhead Regional Medical

Center, where doctors cracked the suspect's chest and did a heart massage to save his life.

Three months later, the San Bernardino County District Attorney's Office called to say they were closing the case, as medical costs had shot up to over $300,000, including daily dialysis. The district attorney said, "You pretty much dealt out more discipline than he's ever going to get, and because he wasn't armed when he broke into your house, all we have is really a residential burglary. If it's okay with you, we're just going to go ahead and drop charges in the case." My wife thought this was total bullshit, but I got it. A few days later, the guy was discharged with several bullet holes and a colostomy bag. He walked out the front doors of the hospital to God knows where.

Days later, after the shooting, Ricardo, a good teenager who lived across the street from the Rialto house, mentioned he had seen the whole thing. He told me that the only reason he didn't come out to help was because there was another guy standing in the front yard behind a tree in the shadows, just out of view, and when suspect number one stopped banging on the door, Ricardo heard them plotting—arguing—about how to get me outside in order to probably hurt or kill me and my family. At the back end of a large lot next to Ricardo's house was an old, beat-up shed where drug deals went down and general ruffians hung out. They'd had a little bonfire that night, and Ricardo had seen them talking and sort of fighting with each other when one of them had said in Spanish, "Hey, don't fight with us, man. Go fight that cop who lives over there across the street." The knucklehead suspect first went to the wrong house, one north of me, got scared off by a big German shepherd, and finally made it to my door. Turns out, the suspect had been on cocaine and alcohol when he'd tried breaking into my home.

Nine months later, we moved "up the hill" to Victorville in California's high desert (the western region of the Mojave), sharing a cul-de-sac with a deputy sheriff. Hanging around the garage one day, we started talking police stuff, and he told me he had an inmate in his facility who thought he had been shot four times by "Rialto PD" a year or so prior—it was my guy! In an odd twist, this deputy had been at the hospital the night of the Rialto shooting with another inmate on an unrelated arrest when he'd seen the emergency room doctors doing the heart massage. Weird, small world. The suspect was back in custody in Victorville because he had killed someone in a drunk driving accident about a month prior, with alcohol and cocaine in his system, just like in Rialto.

Looking back now, the Rialto incident could have been catastrophic had I not been home that night. This happened in September 2004, at the height of the of the LA County Fair when I would've usually been there working overtime. If I had not crashed on my dirt bike a month earlier and broken my foot, my pregnant wife and toddler son would have been at home, unprotected from this maniac. It's like my old sergeant used to say about turning left or right on Mission: Every decision has the potential to change your life, for better or worse.

Man with an Axe

One hot summer night after my foot was all healed up, I was northside, working an overtime shift on patrol, when a "man with an axe" call came over the radio. You didn't hear that every day, but even so, it was a full moon, so I wasn't surprised. I hustled over there, rolling Code 3 (lights and sirens, high speed) and arrived in a neighborhood to see a large Hispanic male—all of six feet six inches and three hundred pounds—without a shirt, sweating profusely in the middle of the street, waving an axe around and trying to get a cop to shoot him. He was obviously on some type of drugs, and in fact, I recognized him as one of our local gang-family members with a history of some real violence towards police officers.

As the would-be lumberjack ambled around, more officers showed up, about ten of us now, with beanbags and Tasers and lethal cover ready, trying to calm him down. But he kept right on yelling and threatening anyone nearby; it looked like this could turn deadly any second. Radio dispatchers had the air cleared with a tone for a high-stress incident that sent even more units rolling Code 3. Suddenly, a voice with some boss behind it lit up the radios. Sergeant Waltman, a tenured sergeant with legendary street reputation—he had even appeared on multiple episodes of *COPS* in the 80s—said, "Nobody do anything until I get there. Tell Randy (suspect) I'm coming."

Thirty seconds later, Sergeant Waltman roared on scene, jumped out of his patrol unit, lit a cigarette, and walked over to the suspect. By this time, Big Randy had stopped bellowing and waving the axe around like he was flagging down a logging truck. Waltman stared at him for a few seconds, then calmly said, "Randy, put the axe down."

"Hey, Blondie," Randy said. Waltman was known on the street as Blondie due to his reddish-blonde hair and New York accent. "I'm sorry." And just like that, Randy dropped the axe. Waltman talked him straight into handcuffs without flinching. Badass defined.

Birds and Machine Guns

I'd been on the SWAT team for about four years, working my way up from the perimeter to entry team, when the whole crew headed to Laverne for a training session with Foothill SWAT, a multi-agency unit made up of several smaller cities. A special guest speaker, Bob Gallegos, was on the docket. Bob was one of the original guys who'd helped start SWAT in LA back in the seventies. The dude is a straight-up legend, a badass all the way.

A brief look back at SWAT's origins: The recognized "founder" of what we know today as SWAT was John Nelson, a Vietnam War veteran, former marine, and LAPD patrol officer under Daryl Gates in the early 1960s. The Watts riots in 1965 had inspired efforts to form a tactical team to manage crowd control. Gates had tapped Nelson for his military training, and his department's nascent team had gotten its first taste of the "limelight" in an hours-long shootout at the Black

Panthers' southeast LA chapter headquarters in 1969. Five years later, in May of 1974, SWAT had leaped fully into the public eye (on live TV, no less) during another heated gun battle with the Symbionese Liberation Army.[1] Subsequent incidents over the decades had since solidified SWAT units' critical roles in everything from active shooter events to terrorist attacks.

A room of around thirty or forty of us attended the training session in Laverne. Bob gave a lecture about the history of SWAT, sharing war stories, explaining team entries, and discussing the brotherhood that comes with being in SWAT. But there was one guy in the room from Upland PD who kept asking ridiculous questions. You know that one person in every class who asks questions that make everyone shake their heads? He was that guy. After a few dumb questions in a row, Bob finally just snapped and said, "Listen, goddammit, if worms had machine guns, do you think birds would fuck with them?" The whole room cracked up. "You're going to be busy out there. Do what you got to do."

It was the kind of saying that stuck with us and became a mantra for the team, a reminder to stay focused and get through whatever situation we were facing.

After making it onto the entry team, I started going on dynamic entries—high-risk warrant operations where a judge has authorized an entry without prior notice, so it was usually pretty intense. We weren't busting down doors for unpaid parking tickets; most of these were to look for guns, gang members, drugs, homicide suspects, and others of dangerous ilk.

When I started with the team, I met one of the senior guys, Mel, who had eight or ten years of experience on me. He was a quiet, reserved guy but a stone-cold operator. He embodied the quintessential SoCal street-cop—impeccable uniform, top-shelf physical condition, meals of Copenhagen and black coffee. He had a quietly confident swagger earned from years of taking down bad guys—there was a presence about him. I'll never forget my first dynamic entry with him. We were about a minute out, and he said over the radio, "Hey, guys, I've got a really bad feeling about this one." And I'm thinking, *What the hell? Who says that before a mission?* But that was the tradition among the veterans, who had an odd ability to calm our nerves with reverse psychology before things got real.

Eventually, Mel retired, and I moved into the breacher position. We continued the "bad feeling" ritual for a while until Dave Estrada (no relation to Erik) and I started a new tradition of our own. I'd gone through SWAT school with Dave, one of the finest human beings I've ever known and an even better friend. We both loved Kenny Rogers' *The Gambler* and decided that would be the new refrain. Pulling away from the office in full battle dress at three in the morning to hit a house full of bad guys, we'd sing like Kenny to dilute some of the tension.

Mel was also a GTA monster. He'd had street pursuits almost daily as a new cop, many of them grand theft auto, or "rolling stolens." The department gave out little pins for these, like mission marks on old warplanes. Mel must have had at least fifteen 10851 pins (named for California's stolen vehicle code). He was the master, and I think they stopped giving out the pins because he had so many of them. Mel was always willing to help you out with anything you needed and was a calming voice in even the most critical of incidents. He was definitely the last of the old breed.

Hard to Unsee

A day in the life of a police officer is largely hidden from the general public's eye—and most definitely its perception. It is far beyond simply giving out speeding tickets; in fact, many calls involve saving lives, as police units almost always arrive several minutes before ambulance or fire department crews. Early arrival means you see things most people don't see and don't want to see. Things abhorrent, maddening, remorseless, frightening, and everything in between that sear into your memory, often remaining long after you hang up your badge.

I got called one night for a fight in progress at Pomona's Fox Theater, a restored movie palace from the Golden Age of Hollywood in the forties. Today, they hold different kinds of shows and concerts, always attracting big, lively crowds. I rolled up, and people, mainly younger kids eighteen- to twenty-year-olds, were running all over the place, mostly away from the theater. Except for one kid who was sitting on the sidewalk with his knees up and ankles crossed, sort of criss-cross applesauce. His right hand was pressed against the front of his neck. I rushed over and asked if he was okay. He looked up at me with a face flushed with fear, shaking his head back and forth.

Another officer arrived, and we gently pulled the boy's hand away, revealing a clean, deep gash, like a surgeon had been at work, and blood pouring out in a hurry. We could see clear to the back of his throat. Ghastly. I pressed my hand back to his neck and said, "Don't move. Don't move. Stay still. Help is coming."

I heard sirens fast approaching as the kid looked directly at me, scared to death of how bad this was. He couldn't speak, but his mouth shaped the words: "I don't want to die."

We sat there for two or three minutes, waiting for the paramedics. Eternity. I tried to think of something to say to reassure him. I silently prayed instead. A medic finally ran over, calling, "What do you have?"

"It's bad. A deep vertical cut straight through his throat." The medic team took over and put the kid on a gurney, loading him into the ambulance. I watched them drive away, hoping, but found out later the boy had died. He was fifteen years old.

What had led to the tragedy was a mess. As the concert ended, two guys got into a shoving match. This young kid, just a bystander, stepped in to try to break it up, do the right thing. The suspect, who we believe was an Asian male, pulled out a straight razor, probably with the intention of slashing the bystander's face. Instead, it caught him perfectly on the neck, slicing it wide open, and that was it. He'd been gone only minutes later.

I'll never forget when the boy's mom showed up and I had to tell her that her child was gone. It's the hardest thing to do.

A similarly unsettling day on the job involved a brain on a bed. I was in training, still with my field training officer (FTO), so this situation really threw me straight into the deep end. We motored up to an apartment complex on the north end of town, found the right place, and walked in through the open front door. The mother of the victim sat on a beige recliner, calm as can be, no tears. A gunpowder smell hung in the air, but we didn't see a body in the main room. I asked where it had happened, and the lady just pointed toward the back bedroom. We walked down the hall to a huge bedroom with a king-size bed in the center and a deceased man and a 12-gauge shotgun on the floor next to it.

He had sat on the edge of the bed, placed the gun on the ground, and jammed the trigger with his thumb. The barrels were in his mouth, and the blast had blown off the entire top of his head, which was all now splattered on the wall behind him. There were no pictures or anything to block the mess—just a plain white wall with a perfect blast pattern. In the middle of the pattern, I could see a perfect imprint of his brain, the veins and valleys and everything. I had a split-second, oddly timed flashback to when I'd been a kid, using Silly Putty to copy the Sunday comics and paste the images into my school notebooks. This was far less entertaining. The gentleman's brain had bounced off the wall and landed in the middle of the bed, just lying there like yesterday's laundry, while his body sat crumpled on the floor, a bloody, grotesque mess. I remember thinking, *Well, that's how my day's going to start.*

We sealed off the scene, my FTO declaring the guy dead at X time, and stepped away to wait for the coroner. (There was clearly no need to transport him to a hospital or verify he was really dead.) The coroner investigator was coming from downtown LA, usually a two-hour drive to anywhere, so we had a little time to spare, and this is when cops develop their dark sense of humor. If you don't disassociate yourself somehow, you'll go crazy.

Then, seemingly from out of nowhere, in walked the lead coroner investigator, a woman with shoulder-length red hair and bright lipstick, wearing a dress and fishnet stockings under her coroner jacket. I certainly wasn't expecting that. Even a woman in that role at the time was uncommon. Most of the time, in fact, coroner investigators oddly resembled their cadavers of the day. But she was cool and casual, like she was strolling into a coffee shop instead of a crime scene. She asked questions about the body and what we'd done so far—whether we've touched anything, emptied the pockets, etc.—and then did her thing.

At one point, she turned from the body, now prostrate on the floor, and asked, "Hey, can you do me a favor? Go grab me a roll of paper towels?" No problem. I thought maybe she'd stepped in something. I handed her the towels, and she rolled the body over and fished his wallet out of his back pocket. "Can you pull the driver's license out for me?" She looked closer at his head, as if critiquing something she'd just painted, and asked for a few more paper towels.

It got weirder: She lifted up the front part of his face, basically a rubber mask with nothing left inside, and stuffed more paper towels into his cranium to try and make it look more like a face. I stood there, completely dumbfounded, until she said, "Okay, hold that ID up closer to the head. Let's make sure it's him." Apparently this was the best way to accurately identify a body? About ten minutes later, her crew showed up—the guys who did the heavy lifting—and zipped the corpse into a body bag. All done.

She handed me a case number and said, "All right, have a good day." Just like that, she walked out, and I was left standing there, still processing what had just happened. That whole situation was just bonkers. And when I started telling the story to people later, everyone knew her. "Oh, yeah, that's her, she's crazy," they'd say. Turns out, she was known for wearing those fishnet stockings, but that was her style, and the blood and guts were just part of the job. You never knew what you'd see next.

That was not the last time I saw that particular investigator, either. At another dead body call, the deceased had been dead for three or four days and was bloated, with fluids seeping from every orifice. The unmistakable decomposition smell was absolutely horrible. Though the situation overall was relatively uneventful—just a usual day finding a dead person—she asked for my assistance. I was putting on two layers of latex gloves when she laughed and remarked, "He's not going to bite you." *Yes, I know, but give me a minute. It's disgusting*, I thought. How she dealt with death so gracefully, I would never know.

The scene reminded me of seeing my first autopsy during field training. As part of the program, we'd spent a week rotating through units in the detective bureau—one day with property crimes, another with homicide, and so on for four consecutive days. I was assigned to shadow Detective Spencer, who was working homicide at the time. He asked if I'd ever seen an autopsy.

"No, sir," I replied.

"All right, well, you're going to see one today."

We drove down to LA, a little over an hour from Pomona, but before heading to the medical examiner's office, we stopped at Tommy's Chili Burgers—a downtown staple. Spencer insisted we grab some food there. "Come on, we got to get this chili burger," he said. I remember how delicious it was. I gobbled that thing, completely unaware that he was setting me up.

After finishing our meal, Spencer casually mentioned, "We just got to run by the autopsy real quick. They're doing our homicide case."

The examination room looked exactly like in movies and TV shows—a large space with four tables on each side, coolers for storing bodies, and a distinctive smell. Not the odor of decomposition, but a cold, clinical smell that's difficult to describe. I immediately started feeling uneasy, and we hadn't seen any bodies yet.

We met with the coroner, took notes, gathered information, and watched from four feet away as the autopsy began. The coroner pulled back the sheet and started by outlining the chest plate with a scalpel. Then he took what looked like pruning shears and cut through the clavicles and ribs—*clack, snip! Clack, snip!*—before lifting off the entire chest plate to expose the organs.

It was bizarre watching him manipulate organs, taking measurements and noting observations. I wasn't feeling well but internally vowed not to cave in. After examining the torso, the coroner moved to the cranium, cutting all the way around it with the equivalent of a Dremel tool, and removed the top of the man's head. *Schloooop, pop!*

I couldn't fight it anymore and sprinted to the bathroom to let loose my lunchtime chili burger. Spencer busted up laughing, but when we returned to the department, he didn't tell anyone about it. Sadly, he'd taken his own life a few months later. It's truly sad what this job will do to some people and what demons they'll have to deal with. You just never know what your partners are going through.

Yet another suicide event, toward the end of my career, again involved a shotgun and a brain. Dispatch sent us to Park Avenue on Pomona's south side to assess the deceased and the associated scene. When we arrived, a teenage girl—probably between twelve and fifteen years old—stood in the front yard, covering her face with her hands, crying uncontrollably.

"Where is he?" we asked. She just pointed toward the garage, a typical working-class gangster hideout turned crash pad—ratty couches, deteriorating chairs, an old TV, a weight bench, N.W.A and Oakland Raiders posters on the walls. The father of the teenage daughter sat in his easy chair, wearing distinctive slip-on corduroy loafer slippers, high socks pulled up his calves. He had on long Dickies shorts, and one hand was visible on the armrest.

As we approached from behind, guns raised, I could initially only see his feet and part of his profile. Coming around farther, I spotted a shotgun positioned between his legs, pointing up toward him. Moving forward, I saw he wore a Raiders jersey with a Mr. T bling starter kit around his neck. He had put the shotgun in his mouth and pulled the trigger, the blast rocketing through his ear and the top of his head, exiting sideways.

He remained sitting there in the chair, and from a casual glance, he looked like he'd just fallen asleep. But his skull fragments and brain matter were splattered all over the *Straight Outta Compton* and Raiders artwork on the wall behind him. Later, I learned the gentleman had spent the past eight years in prison, missing most of his daughter's childhood. Recently released, he'd returned home to a wife who wanted nothing more to do with him. He couldn't do it any longer.

Dead Body Stinkers and Six-Inch Cars

There's nothing else like the smell of a decomposing human. If you've encountered it, you know—human decay has its own distinct odor that makes itself known long before reaching the source. We got a call one day, down on the south side again, to a converted garage in someone's backyard where an older gentleman lived, renting from the family in the main house. They reported not seeing him for two weeks, unusual since he typically went for walks every couple of days. And now his place was really starting to *reek*.

When we arrived, one of our department veterans walked up with a tub of Vicks, dabbing some under his nose. He offered up the tub, "Want some?"

"Nah, man. It can't be that bad, ha!" I replied, trying to play tough guy.

He just laughed and said, "All right, good luck in there."

As we approached the little "house" out back, I noticed sets of three-by-three windows on either side of the front door with what looked like a dark tint. *Well, maybe the guy is trying to cut down on the electric bill.*

But when we opened the door, the smell hit like a truck. You can't turn around and leave, so all you can do is put a hand to your mouth and push forward. Inside, we found the man completely bloated and distorted, his eyes and mouth teeming with maggots. (I should've taken the Vicks!) Then I realized what I'd mistaken for a window tint was *flies*—thousands of them completely blanketing the windows from the inside.

When the coroner's team arrived for recovery, I said, "I'm sorry, guys. I can't help with this one. I'll be outside." Horrific.

High-speed traffic accidents create another category of unforgettable scenes. One particular incident involved a young man and his girlfriend racing southbound on White Avenue in a Mitsubishi Eclipse. The driver lost control and slammed into a huge traffic pole at an intersection, hitting it directly on the driver's side at just shy of one hundred miles per hour.

The impact compressed the vehicle's width—normally about five and a half feet from the driver's door to the passenger door—down to just six inches. Within that tiny space was the driver's door, center console, passenger door, and both occupants, all catastrophically compacted.

Looking at something like that haunts you, but you still have to do the job. You can't say, "Hey, time out, everyone. I need to take a break for a minute." You take a deep breath and work through it. Protect and serve, no matter what.

CHAPTER FOUR

THE WORST DAY

Two Barbies and Leg Irons

As if normal policing and my life hadn't been chaotic enough, after the state 148 trial was dismissed for reasonable doubt, we heard rumors that the FBI was getting involved. My internal affairs case was "tolled"—meaning they paused the internal investigation. They couldn't sign off on it because of federal involvement. Only after everything was resolved would they actually close the case and put it all to bed.

By this point, the FBI was interviewing dozens of officers and witnesses—all kinds of craziness. People were telling me, "Yeah, man, I had to go sit with the grand jury in downtown LA and tell them what happened the night of the fair." I was just getting little tidbits here and there, stuck waiting in fear and uncertainty of what would happen next.

The FBI's allotted investigation time was from May of 2016 to October of 2017, during which they concocted a story that I'd been "hired out" by one of Carlos Alvarez's classmates at Bonita High School, Vega, to find Carlos and rough him up at the county fair because of an argument they'd had in class. Of course, that was completely false.

At the time, I was still working in the training division as a background investigator and as an advisor for the Pomona Police

Explorer Program, which brought in high school–age kids once a week to participate in physical fitness activities, drill training, and police scenario simulations. The Explorers could then go on to compete against other programs in a yearly competition in Las Vegas.

Vega was a solid, upstanding kid, and Carlos was, well, not. They had nothing in common, no issues with each other, no fights. In fact, they didn't even talk. But someone in the class had made a comment about the fair, and somehow, the FBI got wind of it and decided to go down that rabbit hole.

They interviewed all thirty kids from that class, and guess what? Not a single one of them had ever even seen Vega and Carlos speak to each other, let alone heard anything about this story the FBI was pushing, so they were forced to drop the Vega angle. But the Feds kept on, going as far as subpoenaing forty thousand alcohol receipts from the fair that day and asking for every credit card statement from anyone who bought alcohol. Later, we would learn that the FBI and DOJ subpoenaed these credit card statements for one reason: to create a thirty-thousand-page mountain of discovery to drop on our defense team two days before trial. It was ridiculous—a complete fabrication to create a crime, spin a narrative, make it somewhat believable, and start building a case around it.

That was why it was so shocking that they got an indictment. But did they?

Around August of 2017, the FBI called to interview me directly. One of our internal affairs liaisons—we had someone who coordinated between our department and the FBI—contacted me and said, "Hey, the FBI wants you to sit down in front of the grand jury. You're hearing about all these different people testifying; now they want to hear from you."

I felt I had nothing to hide. "All right, let's set a date."

He suggested I call my attorney first, so I got in touch with my PORAC Legal Defense Fund attorney, Michael Schwartz, who had represented me on some prior IA investigations. This defense fund proved to be invaluable to me and my freedom. I paid six dollars every paycheck for this legal coverage, the cost of which ultimately ballooned to over one million dollars as the case got more serious. I knew Michael very well and told him what was going on, and he immediately said, "Stay there. Do not answer another phone call. Do not talk to another person. Wait for me to get there."

I was working at the training division, an off-site location miles away from the police department. Michael met me there, sat down, and explained, "Okay, I'm going to tell you what's up with the FBI and the DOJ. Respectfully, we don't talk to the Feds, and I'll tell you why. Because when you go in and sit down, you have no right to legal counsel or representation. You don't get to have your attorney, you don't get to have it audio recorded or video recorded. There's nothing capturing real-time testimony that may be used to protect you."

He was adamant: "You do not ever go in and sit and talk to the Feds, period."

I couldn't believe it. I said, "What are you talking about? If I get a radio call for a shoplifter at Walmart, I've got to video and audio record everything."

"Because the Feds want to control the narrative. That's it. So you're not talking to them."

"Michael, I want to put this thing to bed. If the FBI just wants to talk to me for ten minutes—"

"If you talk to the FBI or the DOJ or the grand jury, I will not represent you." Damn. Well, apparently, he was serious, and since he was one of the best in the business in California, I took his word for it.

We respectfully declined a conversation based on the Fifth Amendment—the right against self-incrimination. Things kind of floated away after that. Michael and I had a couple of conversations over the next few weeks, but he hadn't heard anything new. Everything seemed like it might just fizzle out.

It didn't. Driving home from work on Wednesday, October 25, 2017, my phone rang as I exited the freeway over to Oak Hill Road. It was Schwartz.

"Hey, what's up, Michael?"

"Where are you?" he asked, with more than a little angst.

"About two minutes from home. I just got off the exit."

"Okay, call me when you get home."

"No, no, I don't play that game. Tell me what happened."

"They indicted you," he said.

"What do you mean, they indicted me? What does that mean?"

"That means that you have to go down to Roybal [Federal Building] tomorrow and surrender yourself to the FBI. You'll be arrested, booked, processed, and arraigned."

"What?" All the blood rushed out of my body.

He said, "Call me when you get home."

I parked in our driveway, walked in through the garage, collapsed onto the ground in the hallway, and started crying, hard. My wife heard me and rushed over. "What's the matter? What's wrong?" I couldn't get any words out. My daughter and son ran to me with concern, and soon, we were all huddled and crying together.

I gathered my composure and gave them a quick explanation of what Michael had told me. They didn't really understand what that meant either. We were all in disbelief and needed answers. After a few minutes, I called Michael back to find out what we were facing.

He first related that he was so glad I'd answered the phone. The FBI was assembling a tactical team to come to my house and serve an arrest warrant on me, related to the indictment. Michael went on to explain that he'd gotten wind of this by chance and called the FBI agent in charge. He'd pleaded with the agent to let him get me to turn myself in, and the agent had relented, emphasizing that if I was not at the Federal Building in LA at nine the next morning, they would serve the arrest warrant at my residence. This was absolutely insane. Knowing common-practice operations, I firmly believe they were planning to hit my house with a full-on tactical team, armored vehicles, multiple agents, FBI raid jackets, flashing lights, bullhorns, and the whole nine yards like I was a cartel boss. I couldn't help but think about how they wanted me to do the walk of shame out my front door in handcuffs at gunpoint, all in front of all my neighbors and kids.

He got straight to the point. "Tomorrow will be the worst day of your life. Hands down. This will be the worst day of your life."

I could hardly breathe. "Okay, thanks for the context. What do I do?"

"Number one, all firearms and ammunition inside your house need to get out immediately. Call a neighbor, preferably someone who has a safe, and remove all firearms and ammunition," he started. "Tomorrow morning, you need to be at Roybal Federal Plaza downtown at nine in the morning. I'll meet you there, we'll connect with [FBI agents] Raye and Lane, and you'll surrender yourself. They're going to handcuff you, take you in the back, and book and process you the same way they do any other criminal."

I didn't like any of this but said, "Yeah, that sounds kind of unnecessary, but I get it."

He added, "Do not bring anybody else with you except your wife."

"Can I bring my oldest son?"

"Yes, but no interaction. They have to stay where we can see them. They don't get to talk to anybody, nothing. Also, don't wear shoes with laces, don't wear a belt, no hats, no jewelry, because you can't have anything like that in lock-up."

That night, all my neighbors and kids came over. I was completely gutted; how did one even begin to comprehend what had just happened? My neighbor Steve took my guns and ammo to his house and put them in a huge safe in his garage, and we all leaned into each other's hearts and strength.

Naturally, we didn't get much sleep that night but got up early the next morning. It was normally a two-and-a-half-hour drive from my house to downtown LA, let alone on a Friday morning during rush hour, so we left around four o'clock in the morning—myself, Amiee, and my oldest son, Blake. Unbelievably, as if this day wasn't going to be bad enough, we got a flat tire on the way! Fortunately, it was a slow leak, and I noticed it before we got on the freeway. I turned around, the car limping back to the house, where we changed vehicles, then took off again.

We arrived at the Federal Building around 7:45 a.m., well ahead of schedule, and parked in the outdoor lot across the street, watching all manner of attorney types, DOJ officials, and FBI agents walking in.

We followed suit around 8:30 a.m., passing through the magnetometers and security searches to find ourselves in a humongous, domed space with ornate columns, soaring windows, and extravagant architecture all around. I felt like I'd just walked into St. Peter's Cathedral. It was true that this building would be where my judgment was passed down, though not from God. Several hallways shot off in different directions from the main concourse. We picked a bench seat in the main room and settled in to await what would happen next.

A few minutes later, Michael arrived, dressed in his traditional Orthodox Jew attire—yarmulke and long, black-and-white suit. We discussed what was going to happen, and he reminded me to expect certain treatment that I was not accustomed to. He said to Amiee, "I'll call you as soon as I hear anything, but don't be surprised if you don't hear anything until three or four o'clock this afternoon."

I gave Amiee a hug and kiss and hugged my son, then walked with Michael over to an information kiosk in the center of the room to meet the two Barbies. FBI Agents Raye and Lane looked like they'd just stepped off the set of a TV show, each weighing in at maybe 102 pounds soaking wet, wearing their little power suits.

Michael made the introductions, and Barbie Number One asked, "Did you prep him about what's going to occur right now?"

Michael said, "Yes, Chad understands the process."

The Worst Day

I spoke up: "I have no problem with you doing your job. Do whatever you need to do." I thought they would be respectful, but I was sorely mistaken. The Barbies waved over two more agents, their "Ken doll" counterparts in matching skinny suits. I felt underdressed in just a white dress shirt, slacks, and slip-on shoes. The Kens put me in handcuffs behind my back and walked me down a hallway to a T-intersection, adjacent to the women's restroom. Across from that restroom was a small alcove where it looked like there might have been a soda machine. They immediately made a 180-degree turn, shoved me against the wall in that alcove—it wasn't super deep, maybe a foot—and started searching me. "Spread your feet," they ordered, doing the whole "rubber glove" search up and down my body. Then one of them reached for the front of my pants, undid my buttons, unzipped my pants, reached his latex-gloved hand down under my nuts, doing a full crevice check. *You've got to be shitting me.* I said, "That's not necessary, but okay. I'm not going to argue with you." I was clearly at a disadvantage here, so I just tried to play it cool.

Mind you, this whole spectacle happened right outside the women's restroom—more than a few ladies walked in and out, shooting me looks like they were wondering what the hell was going on with this guy. Finally, the agent redressed me, and we headed down a couple more nondescript white hallways to a metal door with a camera and a door button. They hit the button, a male voice answered from some secret place, and we walked in.

Immediate change of scenery. We were now in a booking area operated by the federal marshals—they handle all in-custodies for the Feds. We kept walking, and the clean white hallways turned into concrete floors and dull gray walls. On the right side were six or seven small visitation rooms with steel doors and little windows where inmates would visit with people through the glass, talking on phones like you see on TV shows.

We walked past those to a main reception area that handles fingerprinting and photos, with a couple of small stainless-steel picnic tables nearby—though these were decidedly very un-picnic-like. To the left and forward, I saw day holding cells and next to those, a glass bubble containing all the action—buttons, security monitors, serious people with clipboards, and two marshals dressed in green jumpsuits, apparently expecting me.

One of them took hold of me, turned me around, and said, "Hey, you know the drill—you're a police officer, right?"

"Yes, sir. Yes, sir. I got no problem. Do whatever we need to do."

He took the cuffs off and said, "Put your hands on the wall for me. Go ahead, take your shirt off." I handed it to him. "Take your trousers off, take your socks off and everything down to your underwear, and hand them to me." He searched the items for contraband as he would with any other inmate.

I knew what was coming next. Guess he was saving the best—the ass-crack check—for last. I appreciated that he wasn't a jerk like the Kens, but it was indeed humbling and embarrassing.

It was brutal but at least quick, if that's any consolation. Afterward, he handed my shirt and pants back. It was the second time in the span of an hour that complete strangers had gotten intimate with me.

The same marshal politely said, "Do me a favor, step over here," signaling to the fingerprint area to my left. I did so and sensed an immediate change in personalities compared to the FBI guys. The FBI agents had a fake toughness about them—they'd needed to have my hands crimped when walking me, all the while exuding an air of superiority. I wasn't a high-profile, most-wanted-list fugitive; I'd driven in, flat tire and all, to turn myself in.

The marshals were totally cool. Nobody put their hands on me or did anything overly macho. I stepped to the print module, and a marshal got all my prints while the two FBI goons stood there watching. After taking my prints and picture, he directed me to one of those benches facing the wall. The FBI agents took post just to my right at the edge of the table, hovering like the robotic bodyguards of a cartel boss.

A felony booking requires a DNA swab, which is a simple procedure for most, but the two FBI goofs were like Keystone Cops. One read the directions on how to do a swab while the other tried to open the packet with oversized latex gloves on his small hands, completely lost. The marshal, watching the fumbling from across the room, finally walked over, yanked the swab packet from Goof Number One's hands, and said, "I got this. Don't worry about it." Just like any other street cop would do, he ripped it open and did the swab on my inner cheeks. "Okay, one more. Thanks, man. Appreciate it." He signed the form and handed it back to the FBI guys, who were visibly embarrassed and generally pissed off by what had just happened. Too bad their arrogance was all for show!

The marshal escorted me to the day room—a twenty-by-twenty space with a solid bench and a toilet in the corner—where I saw Neaderbaomer being escorted into the booking area by two other agents, preparing to put him through the same routine. My two FBI agents started muttering to each other under their breath, and as we passed, one of them exclaimed, "No contact for these guys. No contact."

The marshals looked at each other, confused, and I said, "Hey, man, we've been working together for the last two years. What do you mean 'no contact'?"

The marshal gave me that tight-lip sign like, *Hey, stop talking. Relax for a second. I got you. Just give me a chance.* That was the same thing I would do for somebody if the roles were reversed, given that I was now the one in the cell. I didn't say another word.

The lead green jumpsuit emerged from the bubble and talked to the FBI dynamic duo. After some visible disagreement, the goons headed back down the corridor while the marshals discussed something among themselves. Finally, my main guy stepped back over, shaking his head in an *I'm sorry* way. Uh-oh.

They put me in waist chains that go around you, with each handcuff hooked to the chain at your belly. That chain then extends down your backside to leg irons. All secured, a marshal brought me to one of the little interview rooms. I shuffle-stepped my way there, chains going *clink, rattle, snap!* just like you see in the movies. He opened the door, and I nearly ran into a plexiglass wall. The room was four-by-four feet, occupied only by a counter, phone, and metal stool with a big loop at the bottom.

"All right, hey, do me a favor, man. Have a seat right there."

I sat down, hands still in front of me at my belly, and he secured the waist chain to the metal loop. He took the leg irons off, but now my waist was chained to the base of the stool, so I couldn't stand up more than about six inches.

"All right, man, I'll be back with you in a bit." There I sat, locked up like a common thug, sinking further into disbelief and denial. By this time, it was about 9:15 a.m.—four hours earlier, I'd been dealing with a flat tire in my own car in my own neighborhood.

Through a reflection on the glass, I saw someone go by—it was Prince, my partner the night of the fair. It wasn't just me they were going after; it was everyone. Ten minutes later, some kind of blasted-out, tattooed guy in an orange jumpsuit—a freaking MS-13 looking gangster—showed up in the room next to me. His attorney or somebody talked with him for a minute, then left. The brute took a moment to glare at me, then sat there like a statue. A gut punch of reality hit me all of a sudden. What the hell was happening? How was I in here with this guy?

Pretty soon, the gangster exited his cubby hole with a marshal, and Neaderbaomer took his place. He was a friendly face, someone I knew. A glimmer of hope? But he didn't even glance my way; in fact, he kind of turned his back to me. Why was he doing that?

It got even more interesting when a marshal brought Prince into the room next to Neaderbaomer. There we were, the Three Musketeers in a tidy little row. Prince and I shared a head nod. *I really wish I could talk to you right now, brother*, I tried to convey.

About 11:30 a.m., the marshal brought in some food.

"Thanks, but I'm not hungry," I said.

"Well, I got you some lunch." He set down a six-inch Subway sandwich and a bottle of water on the counter. I still had my hands in waist chains and had to crunch over to try to unwrap and eat the sandwich. Have you ever tried to open a bottle of water with your hands attached to your belt? I took a couple of bites of the sandwich and left the rest.

After fifteen minutes of dead quiet, the marshal returned, cleared off the counter, and quietly said, "Hey, do me a favor. Keep your hands down where they're supposed to be, but I'm going to unlock that left one. Just don't be waving your hands around."

"Cool, man, no problem."

"Just kind of keep it at your waist so it doesn't look obvious."

"Hey, man, I appreciate that. Thank you very much." A brief reprieve from the shackles felt great.

Thirty minutes after that, a younger Hispanic guy, probably mid-twenties, who looked college educated, walked in and sat down across from me, introducing himself. "I'm your pretrial federal probation officer. It's my understanding that you want to make bail."

Simply hearing those words was another gut punch. How had I ended up on this side of a statement like that? What the hell was my bail going to be?

"Well, we've done the research, and you have enough equity in your home. Your wife is just going to need to sign a fifty-thousand-dollar bond for you." He took some information, running a brief employment history, then asked, "Is your wife going to answer if I call her?"

"Well, she doesn't answer calls she doesn't know. You'll probably want to call my attorney first." I think he went ahead and called her anyway; she answered, and they arranged to meet to fill out paperwork for bail.

I was left alone again, still chained to the stool, until around one in the afternoon, when a marshal I hadn't seen before came in. "Hey, you got to use the restroom?"

"No, I'm okay."

"All right, come on."

"No. Really, I'm good."

"Nope, come on."

He unchained me from the stool but buckled my left hand back up and refastened the leg irons. I shuffled down to the day room, where there was a toilet with a little concrete wall for privacy. In a whisper, he said, "Do me a favor—go in there and go to the bathroom and come back. Make it look like you're going."

"Okay, got it."

"Just relax and sit down. I'll be right back." It was his way of giving me a break. He took my irons off completely so I could sit in the "restroom" for a while with my back straight and get the blood flowing in my hands again. I perched there for about ten minutes—the best fake crap I've ever had. Then he signaled me over to the gate, put the irons back on again, and stepped me off into a small corner.

"Hey, you're Pomona PD, right?"

"Yes, sir, I am."

"You know Dave Estrada?"

"Yeah, I know Dave—he's one of my best friends. We went to SWAT school together."

"Yeah, I used to work with him," he said, "but now I'm here at the marshals."

In the back of my mind, I had that little guy spinning around and saying, "Think about what you say before you say it, even if you think you know somebody." So I asked him a couple of verification questions to double-check his information: "Where was Dave living when you guys met?" "What department was that again?" "What was his wife's name?"

Once I was satisfied that he knew Estrada too, he said, "Let me ask you a question, man. Did you guys bury a body or something?"

"What the fuck? No, dude. This is something that happened at the fair—a non-injury use of force two years ago at the LA County

Fair. It's a big nothing. I have no idea why I'm here. This is absolutely crazy."

"Well, something is weird, man, because listen—last month we had Tanaka in here." Tanaka was the LA County undersheriff, known to be dirty. For example, one time, the FBI had planted informants in the county jail, but Tanaka found out they were informants and moved them to different places to hide them from the FBI. The dude was totally corrupt and ended up going to prison.

"He never came back to the booking area, though. *He* didn't even get treated like this. Somebody is sending you and your partner a signal. Keep your chin up, brother." He led me back to my tiny room to think about that little nugget for a while. Things were getting weirder—and scarier—by the minute.

How Do You Plead?

After what seemed like an eternity confined to my little four-by-four box, the marshals came in and escorted all three of us down a nondescript, cold hallway, joining a line of six rough-and-tumble gents in orange jumpsuits, including my gangster-looking "friend" from earlier in the day. We were an odd bunch, to be sure—six guys in orange jumpsuits and three white dress shirts and slacks.

We all filed into the courtroom together, and I saw Amiee and my son in the audience. This was another major gut punch moment—there I was, shackled in an open courtroom with inmates on my left and right for my own arraignment, all in front of my family. The reality and gravity of this situation were numbing. The judge spoke up and started calling out names: "John Smith, you are charged with X; how do you plead? Juan Valdez, you are charged with X; how do you plead?" And on down the line. My attorney sat nearby and whispered, "Obviously, not guilty. Just say 'not guilty' for each charge."

When my turn came, the judge's voice sounded distant and echoey, like an actor's internal dialogue in the movies: "Chad Jensen, count one: abuse under the color of authority. How do you plead?"

I replied, "Not guilty, Your Honor." We went through all the charges like that, and then it was over.

Afterward, they shuffled us back to the holding area. The guys in orange jumpsuits took a right, and the three of us were led to the left, back toward the booking area where we'd had our photos taken earlier. The marshals chatted for a bit, then walked us into another day room, similar in size to the one we'd been in before, just in a different location. They took off our leg irons and chains, then directed us inside. "It'll be about an hour before you're out of here. Just hang tight."

Michael sat next to me, while Prince was on a bench across the room. I tried to talk to Michael, but he quickly put his finger to his lips and shook his head, signaling me to stay quiet. I muttered, "This is fucking bullshit," and he just gave me a look. Prince smiled, and in that moment, it was clear that things were so messed up, it almost seemed funny.

We sat there for a while until the marshals came back to tell us they were ready to let us go. "Your paperwork will be with your attorney." I was handed a piece of paper with the name and address, in Riverside, of the federal probation officer I'd need to report to. "You need to be there Monday morning at eight. Don't be late, and for no reason can you miss it."

"Yes, sir. No problem."

We were then led out into the lobby area in front of the Federal Building. My wife and son were waiting out there, and it was such a whirlwind and blur, but I think Mike's and Prince's families were too. Everyone stood together in silence on the front sidewalk, and the attorneys from all sides had gathered as well. We all spoke briefly—though it wasn't at all what I'd expected.

Michael (my attorney) reiterated, "Like I said, this was going to be the worst day of your life."

"Yeah, but they didn't have to make it like that. That was ridiculous."

The Worst Day

Neaderbaomer was equally frustrated. "That whole thing? Getting paraded and searched like that in front of the ladies' room in the lobby? Completely unacceptable." He was right. They could've done that search discreetly right outside the entry door, with no one around. It indeed felt like they'd been trying to make a point. We lingered a few minutes, catching our breath, sifting through emotions, and then went our separate ways.

I ended up talking to Neaderbaomer later about this indictment because I just couldn't wrap my head around how they'd managed to get it. I kept thinking, *There must be an indictment since we got arrested, right?*

But he told me, "Well, actually, they didn't get one."

I was confused. "What do you mean, they didn't? We were arrested, so how could they not have an indictment?"

He took a breath and said, "All right, let me tell you how this all went down." He explained that after the case had been dismissed by the state, the FBI had been contacted by Carlos's defense attorney, who claimed that I was lying and that I beat up kids at the fair. So the FBI and DOJ did their investigation—interviewed witnesses, talked to everyone involved. Then they decided to take everything they had from the investigation—witness statements, interviews, and all the evidence—and present it to the grand jury.

The grand jury's job is to review all the evidence, then decide who else to interview and subpoena. The FBI had interviewed around forty people, reviewed hundreds of documents, and even provided transcripts from the state trial to the first grand jury. But after all that, the first grand jury, which also interviewed a dozen or so witnesses, *did not* return an indictment. The DOJ didn't let that stop them. They just picked a second jury. This time, the new grand jury didn't get any of the witnesses or documentation. Instead, they only heard from one person: Special Agent Raye. She walked in and spun a wild story of me yanking kids around at the fair and all sorts of other fantastical lies.

The grand jury listened for an hour, maybe two, and then—without hearing from a single witness, without reviewing any statements or evidence—they returned an indictment. An indictment! Just like that. It was mind-boggling. I couldn't believe it.

And that was how it had happened. It seemed like the whole thing had been a setup from the get-go—like a federal hit job on me. I wasn't alone in this, though. There were three of us caught up in it. Even so, I'd found myself sitting in the defendant's chair in federal court, looking across the room at people who had absolutely lied to get this indictment. They'd done whatever it took to make it happen. No regard for the truth. No shame in how far they'd gone.

I remember telling my attorney at the time, "I just don't get it. I can't believe this is happening. These people are lying, and they're willing to ruin our lives to make this indictment stick."

His reply still echoes with me today: "Chad, this isn't about truth and justice. It's about hanging a cop. That's how you become a federal judge." It had hit me hard, but I knew it was the truth. It wasn't about justice. It was about making a name for people at the top.

The whole thing felt like a joke to them—there was nothing to see here, right? Everyone had lied to get an indictment on my partner and me, and yet no one was being held accountable. It was maddening. It's something I rarely talk about even today because it just gets me riled up. My heart starts racing, so I don't revisit it too often.

The worst part is that in a grand jury, there's nobody defending you. No one speaks on your behalf. And the FBI, the DOJ—they don't even have to provide solid evidence. All they have to do is tell a story, because they're sworn under oath to tell the "truth." But when they lie, nothing happens. No one checks them, and they get away with it. That's the real injustice.

CHAPTER FIVE

GUILTY UNTIL PROVEN INNOCENT

A Horse, Goats, and Depression

On the way home from the arraignment, we stopped by Pomona PD to meet with the chief and deputy chief. Walking up the steps to the front door felt foreign. I'd spent more than twenty years in this place, with the people inside; now, it seemed like I didn't belong. My mind was racing, almost in a daze, and I half expected gatekeepers to block my entrance.

Amiee took my arm, and we walked through the now-ominous portal. In the same corridors and lobbies and anterooms I'd walked since my first days as a police officer, I saw familiar faces, but something was off, as if they were oddly blurred, blocking the connection we'd always had. Some of them offered clipped or awkward greetings; others turned away in a thinly veiled disguise of tending to their work.

We navigated through it all to the chief's office. I had to turn in my gun and badge, a formality that seemed far too brief and void of sentiment for accessories that had been attached to me for so long. I'd worked for the then-chief before; he'd been my squad sergeant, and we'd spent years together on the job. We'd had a good working relationship, same with the deputy chief, so when I walked in, it didn't feel hostile. It felt heavy.

I remember sitting down across from them and feeling the weight of the moment, the gravity, settle over everything. I could see it in their eyes too. They'd found out about the indictment the day before, just as I had. The chief broke the ice. "Well, Chad, you kind of know how this goes. I need you to sign this. It's administrative leave. You'll need to be available to respond to the department 24/7, Monday through Friday. We need your ID, your badge, and your gun." It's common practice for police departments to place officers on leave while investigating serious matters and complaints. They generally continue paying you while they investigate—you just have to stay home and be available to respond to the police station if the internal affairs team needs to speak with you. This keeps you from taking off to Hawaii or some other far-off place in the midst of the investigation.

He laid it out as professionally and calmly as he could, but I heard a hesitant reluctance in his voice. Then came something I didn't expect: "Laura [the city manager] has decided to keep you on the books. Your paycheck stays the same. Your benefits stay the same. There's no interruption."

That surprised me. I figured once the indictment hit, they'd want to distance themselves as fast as possible. But instead, he explained how even my banked time—vacation, sick leave—would keep accruing. At that point, I was already close to the five-hundred-hour cap, maybe 475. I hadn't been using any of that time while on leave, so it was still building. But once you hit that cap, it's use it or lose it—and I couldn't use it.

He said, "Here's what we'll do. The system won't log anything over five hundred, so payroll will keep a manual ledger for you. Every pay period, they'll write down your 11.87 hours of vacation time. When you come back, you can use them."

When you come back.

That part stuck with me. He didn't say *if*; he said *when*. That meant something, especially coming from someone who knew, from the internal investigation, that there was no use-of-force issue here. They didn't know what the charges were really about, but they knew I wasn't dirty.

He finished with, "We look forward to your return."

The deputy chief, Oletti, on the other hand, just sat back in his chair, not saying much. It felt odd. He and I had worked motors together. Our families had camped together, our kids were the same age. But he always kept a line between friendship and work. That was his thing. We'd never grabbed beers just the two of us—it was always a group or family. And that distance was there in that room too.

I turned to him and said, "Well, the chief is retiring in six months, so what's your take on all this?"

He just nodded and said, "I echo his sentiments. When you come back, we'll go from there."

Then we all shook hands, and I walked out, got in the car with Amiee, and drove home.

And just like that, my police powers were officially stripped away. I had walked out of the chief's office that day a regular guy—no more authority, no more uniform, nothing. I hadn't even made it back to the car when the weight of it all really hit me, like the feeling that comes when someone passes away. It's not necessarily the funeral that's the hardest—it's the weeks and months that follow, when everything calms down, the phone calls stop coming, and the daily "Are you okay?" messages fade away. That is when you realize the silence is as heavy as the loss itself.

A big part of working at a police department—like any big organization—is that you get used to the constant buzz, drama, and soap opera stories: Did you hear about that arrest? How about Mitch dating Liz from Records or that dispatcher arguing with so-and-so on the radio? How about that crazy pursuit last night? The dopamine hits keep you energized and engaged, but when that stops, when it all goes quiet, it's like the fire inside you goes out too.

And that's exactly what happened. Within just two or three days, I had no contact with anyone from the department, after nearly twenty years of working with them. It wasn't like they'd been told not to talk to me; they were just afraid to. After everything that had happened and the crazy stories—the FBI tapping phones and seizing text messages, all the rumors flying around—who could blame them? Even my academy buddy, Bill, secretly called my wife's phone from his wife's phone just to talk so we could catch up without "getting caught." Outside of that, everyone else was too afraid to reach out.

The hardest part was the waiting. I was indicted in October 2017, but the trial didn't start until October 2018—thirteen months later. And during that nebulous time, I was essentially already serving a sentence. I couldn't get a job, couldn't go anywhere outside of the Southern California counties. My passport was even taken away. I was basically on house arrest, and it was a lonely isolation—no one wanted to hang out with me. I'd become an exile. The last time I'd felt this outcast was sitting on the Junipero Street curb in Monterey when my stepmom had kicked me out.

Fortunately, I had a small circle of friends outside of law enforcement who really came through for me. My neighbors were a solid support system and were there for me when I needed them. Most of them were not cops, but one neighbor was—my buddy Jeff. We spent a lot of time together, and he was a great sounding board for me. He had a great understanding of law enforcement and the bullshit that goes on since he was from Upland PD. All of them were a blessing to have, especially considering the advice I'd gotten from our senior tactical instructor back in the academy. He'd given us a piece of wisdom that I've never forgotten: "Don't make every friend you have a cop." He'd told us to branch out, to blend our circles. At the time, I hadn't really thought much of it, but looking back now, it made all the difference. I pass that same advice along to my sons now; very few of their friends are police officers or deputies. Nick is a deputy at San Bernardino County Sheriff's Department in California, and Haden is a deputy up in Kootenai County, Idaho.

For thirteen months, everything I did was closely monitored—no police contact, obviously, and no guns or ammo. I couldn't even get a speeding ticket without having to call my pretrial probation officer because even something like that could be a violation, and if that happened, I could be sitting in federal lockup until the trial. When people say, "Oh, they're out on bail," they're not really "out"—it just doesn't look like jail from the outside.

The no-contact order was harsh. It turned out that most of the people I thought were my close friends were just work acquaintances. I'm not throwing stones or anything—that's just life, man. That's how it goes. Every now and then, someone would secretly reach out, and I'd get a chance to talk to them and hear stories, like how the FBI and DOJ had been down at the station, asking for documents on every person I'd ever worked with on the squad, including their full background packets, all use-of-force reports, and, of course, everything from my last five years—arrests, reports, paperwork. We're talking north of ten thousand pages, and the department had to hand it over.

That part was tough. You really do get treated like a leper, and if anyone's caught talking to you, they're putting themselves at risk too. No one wants that smoke.

When I was stuck at home with nothing but time and reflection, I cracked the great book open again, flipping through the Bible and thinking maybe I needed to get back in touch.

Even though I'd been raised Mormon, once I got out of my stepmonster's house, I drifted away from the church. I tried to go here and there, made a couple of friends, but I wasn't really sold. Honestly, a big part of my religion resistance came from how my stepmom had acted—treating me like garbage at home, then putting on a godly face at church. That kind of hypocrisy really turned me off from all organized religion.

So for most of my young adult life, I'd had no connection to church. I still had my personal relationship with God, but nothing formal. No structured religion.

My wife has been Catholic all her life, completed all her sacraments, and always leaned on faith. She asked every now and then if I'd consider taking a few classes to get back into church, maybe complete some sacraments of my own, and look into having my previous marriages annulled so we could have our marriage blessed through the Catholic Church.

Soon, I found myself praying every day and personally needing more, so I made the call and found myself meeting with Father Joaquim, a Catholic priest at the parish in Phelan, California, near our home.

He was from Nigeria, with a thick homeland accent—it took me a little while to fully understand him, but he was just a really good, genuine person and became such a positive force in my life. He set me up with the right sponsors and got me into catechism classes, and by the end, I was baptized, received my First Communion, and was confirmed, and Amiee and I were remarried in the Catholic Church. That whole process of finding faith again was one of the only good things to come out of that entire horrible ordeal. I may not have ever made that leap of faith without it.

Chevy

My boy Chevy. From way back as a little kid, I had always wanted horses. My grandpa and great uncle had owned little prefab "cabins" on a couple of acres each, way out in Yucca Valley, California, essentially the middle-of-nowhere desert at the time. My best friend, Nick, and his family had a cabin down the road as well, and we all rode dirt bikes together on weekends. At my great uncle's place was a little Paint pony named Molly, who was so calm and gentle. We spent a lot of time out there, and I developed a strong affinity for the beautiful animals. Much later in life, that love was part of why we moved up to the high desert to get some land—we settled on two-and-a-half acres—and not long after, I came full circle and got my first horse.

It was my buddy Eli's fault. The Sanchez family lived down the street, and Eli was an incredible horseman and friend. Our kids were about the same age, so our families would hang out a lot, and over time, I started learning the ropes (or the reins) from him—neighborhood cowboy lessons. Eli had plenty of horses and would take me out riding every chance we got. Eventually, he told me, "Man, you should think about getting your own." It didn't take much convincing, and pretty soon, I bought an Appaloosa gelding named Chevy. At the time, he was around ten years old, a former barrel horse—tall, not super stocky, built more for speed. I picked him up sometime in late 2016, just before all my drama in 2017. It was kind of a spur-of-the-moment decision. I didn't put a ton of thought into the cost or the commitment—I just wanted a horse, so I went for it.

But horses aren't cheap. After the initial outlay for Chevy, we built a round pen, a corral, and sourced all the supplies that come with horse ownership, from saddles to bridles. I was still working fifteen, sixteen hours a day, six days a week, and barely had time to ride. I remember wondering if I'd made a big mistake. That horse was pulling a lot of money out of my pocket.

Then the indictment hit, and I'd been placed on leave. By that point, even with limited time together, I'd had a solid relationship with Chevy. I'd been riding, feeling more confident in the saddle. The day after the indictment, Eli came over and said, "Hey, let's go riding." That sounded great, and a casual trail ride turned into one of those life moments. Eli shared some heartfelt stories about losing his dad and how close they'd been. He was struggling with it—dealing with depression, drinking too much, just in a dark place. But he said whenever he got on his horse—Norman—it changed everything. Riding was therapy.

I took that to heart. Maybe it could work for me too. I started riding regularly, every other day, sometimes more. I could ride straight out of my backyard to miles of trails in every direction. Horses have a very calming presence, and that time with Chevy helped me get back in tune with myself emotionally. Even just throwing flakes of hay or alfalfa into the corral and watching him eat made me feel relaxed. I'd spend hours out there brushing him down, saddling up. Sometimes, I wouldn't even ride—I'd just tack him up, then untack. That daily routine became part of my healing.

So I leaned into it even more. We added two mini pigs, then picked up two Nigerian dwarf goats. Those little guys were something else—doing goat parkour, jumping around, just full of life. They brought so much joy. Taking care of them, even mucking their stalls, gave me purpose. When you're used to working all the time, providing for your family, being stripped of that spawns self-doubt and makes you question everything—your worth, your identity, your station in life.

The mind can be extraordinarily relentless, and mine never shuts off. I'd lie in bed at five in the morning, unable to sleep, my brain looping through everything—the arrest, the trial, the time in lockup. It was grinding me up. Some mornings, I'd pray for daylight just so I could get up, grab some coffee, and go sit with Chevy to watch the sunrise. At times like that, I reflected on something my mom had taught me back when I was going through my first divorce. She'd said, "Write 'PMA' on your palm—Positive Mental Attitude. Anytime you feel like the world is crashing down, look at that and shift your mindset." She'd also given me a little plaque with the Serenity Prayer on it:

> *"God, grant me the serenity to accept the things I cannot change, the courage to change the things I can, and the wisdom to know the difference."*

My mom had been a recovering alcoholic at the time, and that AA message always stuck with me. I still keep that plaque on my bathroom counter and look at it every morning. Because there's only so much you can do. You can't stand in the middle of the street shaking your fist at the sky. You just have to accept reality, do everything you can to fight your case, and trust that God's got a plan. You can resist it all day long, but you're not going to change what's meant to be.

Around that time, I ended up talking to a relative of my son's wife, who was in the fire department union. After hearing about what I was going through, he said, "Dude, where is your union? Why aren't they checking in on you weekly? Making sure your bills are paid?" Because as a cop, you live on overtime. Our police officers association, of which I served on the board at one time, was not as strong as it had been in years past. It was mostly used by a select group of ass-kissers and sellouts as a springboard for promotion. I'm sure it's that way at most police and sheriff's departments around the country—guys using their position on the POA Board to gain favor with the command staff. But that's just how it is. That's why they warn you at the academy, "Don't go buy a truck or a boat right away." But of course, we all do. We buy houses, trucks, boats, whatever—because we count on that overtime being there.

Well, obviously, now I wasn't getting any overtime, but thankfully, with Prince, Mike, and me on the books, our paychecks and insurance kept rolling. The newly promoted chief, on the other hand, wanted nothing to do with us; he'd basically marked us with scarlet letters. But the city manager had had the "you're innocent until proven guilty" wherewithal to do the right thing, and I'll always be grateful for that.

So from October 2017 through August 2018, I spent most of my time focused on the animals and trying to stay grounded. Then, in about August of 2018, my attorney called and said, "We need to start prepping. The trial's in October. You need to be completely prepared in case you have to testify." Even though the Fifth Amendment says you are not required to testify in your own defense, it's critical to be prepared. I followed his advice to the letter, spending five or six hours a day going over trial prep at his office in Ontario, California, along with our defense expert, who had testified as an expert at multiple federal trials for similar cases on the use of force, application of force, and defensive tactics.

We were ready. No, I *thought* we were ready. About two weeks before the trial, my attorney called and said, "Have you talked to anybody about this?"

"No, you told me we can't talk to anyone, even co-defendants."

Then he hit me with this: The DOJ had sat down with Prince and offered him a deal—eighteen months' probation if he signed a statement written by the FBI saying I'd coerced him to write his report, that we'd been covering stuff up, and that I was out of control. Prince, to his credit and solid character, told them that wasn't what had happened. They'd pushed hard, telling him he could be looking at twenty-five to forty years in jail. But he wouldn't lie just to save himself. I could not believe what I was hearing! The DOJ and FBI had literally strong-armed Prince and attempted to make him sign a statement that they knew was false. It was a move straight out of a mobster movie. All of this, just to get a conviction.

Up until then, I'd still kind of believed the FBI and DOJ were the good guys. That they'd see the truth, and this would all wash out. I thought it was a misunderstanding—that someone had given them bad info, maybe even fabricated a story, and that it would come out in trial. Then came the Friday before trial. We were prepping everything, figuring out logistics. Schwartz told me to find a hotel and be prepared to spend a month in LA. Saturday morning, he called with another bombshell—true to form, the Feds had dumped those thirty thousand pages of discovery on us at 4:59 p.m. on Friday afternoon. And essentially none of it was related to the case, like alcohol receipts and credit card statements from fair patrons, all meant to bury us. But that's how they work: They outlast you, outspend you, out-resource you. Those tactics give them a roughly 96 percent conviction rate; if I hadn't had legal defense coverage through California's PORAC Legal Defense Fund, I'd have been destroyed financially. My defense ended up costing well over one million dollars between both federal trials. Most people can't survive that financial burden or the relentless accusations.

Unbeknownst to us at the time, buried inside our thirty thousand pages were two FBI 302 forms—statements from Alvarez's brother and cousin saying *he'd* been the one causing problems, not us. They said that he'd been combative, argumentative, and interfering and that the officers had acted appropriately. That was exculpatory evidence—evidence they *had* to share with us under the law. But they'd deliberately hidden it, hoping we wouldn't find it in time, and it had almost worked.

That's the game. That's the system you're up against, designed to defeat you.

CHAPTER SIX

THE FIRST TRIAL

A Courtroom Guardian Angel and Teddy Ruxpin

Federal court in America operates on its own level. Judges, nearly to a one, are anything but neutral and in fact have a deserved reputation of being the third arm of the prosecution. From a defense perspective, we weren't just fighting with just one hand tied behind our backs—it felt more like both hands and a foot.

Going into the first trial, it was supposed to be all three of us—Prince, Mike Neaderbaomer, and me. But Mike's legal team decided that it would be better for him if they bifurcated him to his own separate trial. Mike still wanted to be able to sit in the audience at our trial because he'd been part of all this from day one and had intimate knowledge of all the statements obtained during his internal affairs investigation. His counsel advised against it, however. There was no rule saying he couldn't come, but the advice was to stay out of the courtroom and not draw any unnecessary attention from the DOJ and FBI.

Our strategy for the defense was simple: Defend the truth. Walk through what actually happened. We relied on logic, facts, and our mindset during the incident—all very straightforward, or at least we thought. The Alvarez kid had sustained no injuries during the incident; it wasn't like Rodney King or George Floyd or other cases of outrageous uses of force. It was over in seconds, and he walked away just fine. We genuinely thought it was a nothing case.

But federal court? That's another beast altogether. What they don't tell you upfront is how much of your defense gets chopped down before you even enter the courtroom. Through a process called "motion in limine," the judge essentially decides what the jury can and cannot hear—what's admissible, which experts you can use, what witness testimony is allowed. All typically filed before a trial begins.

Then the prosecution shows their cards. Their entire play was to take testimony from the state-level trial—particularly the 148 charge (resisting arrest)—and redact it. I mean seriously redact it, cherry-picking little pieces that fit their narrative and leaving the rest out. They'd pull out a few words—like "he didn't throw a punch"—and show it to the jury completely out of context to paint it like I was lying and covering the whole thing up. But looking at the full, unredacted statement, it was clear I'd explained everything and never tried to be deceptive. I'd described how, in the fog of the moment, I'd seen a flailing arm coming toward my face. I'd broken all that down in detail. But this abridged version of things made it seem like I had something to hide.

I asked my attorney, "Why can't we bring in the full version? Let the jury see the unredacted copy. Let them read the whole thing in context."

All our defense attorneys, both mine and Prince's, argued tirelessly against the judge's limine order allowing the prosecution to bring in severely redacted written testimony and barring the defense from bringing the full, unredacted testimony from the state trial into court. This was a direct violation of the Rules of Evidence, specifically the Rule of Completeness, which clearly states that if one side, defense or prosecution, brings in a "portion" of a report or out-of-court statement, the opposing side has the right to admit the entire report as evidence for completeness. We couldn't believe that the judge was absolutely violating my rights without a care in the world. He just kept saying that he'd made his ruling, and if we wanted to say anything else, we would have to submit it to the court in writing, which our team did. But we now knew that the jury was not going to see the complete version of the report and get the whole story. It was absolutely crazy that the judge could get away with this.

Basically, the judge was saying federal court didn't work that way. The only way to get that full story in front of the jury was for the defendants to testify.

Insane. I just sat there thinking, *How the hell can the prosecution take redacted pieces of testimony and twist the context, and somehow, that's admissible? But I'm not allowed to bring in the full explanation unless I take the stand? I guess that whole Bill of Rights thing doesn't apply to federal judges.*

Wasn't this supposed to be about finding the truth?

Clearly, it wasn't.

That was when my attorney dropped a line on me that I still think about today: "The only difference between God and a federal judge is that God is not a federal judge." The only person who can hire and fire a federal judge is the president of the United States. Let that sink in for a minute. Talk about unfettered power.

On the Sunday before the trial, we headed down to LA, settling into a four-bedroom apartment downtown. I was lucky enough to have my whole family there this time: my wife; my dad and his new wife, Joyce; and my kids—Blake, Ryder, Sydney, Nick, and Haden—and their wives and girlfriends. We all planned to stay together as a tight-knit support troupe for the duration of the trial.

Monday morning, we filed into the courthouse after standing in a long line outside, making our way through security and the magnetometers. While most government courthouse buildings are older, this one was new—massive, encased in ornate stone, and paved with marble floors. On any other day, I might appreciate such grand architecture, but that day, it evoked the opposite—I felt cold and shivered involuntarily a few times. We took the elevator up to the courtroom, and that was the first time we saw our prosecutors sitting on the other side. I was there with Prince and our attorneys; across the room were the two Barbie FBI agents—Lane and Raye—and the federal civil rights prosecutor, Donald Ash, as well as the lead DOJ federal prosecutor, Lois Frances, the full might and power of the United States Federal Government.

As the trial was set to begin, my attorney, Schwartz, leaned over and said, "Remember to control your emotions, and make sure your family controls theirs. They can't react, no eye-rolling, no gestures. The judge won't tolerate it."

He continued with more coaching, although it sounded more like warnings. "They're going to say terrible things about you. Some of them will be lies. You're not going to get the chance to defend yourself in the moment. You can't flinch, can't show a single expression. Jurors will be watching your every move, waiting for a reaction."

Then came the jury selection. My family sat in the back row, with the other rows filled with eighty to one hundred potential jurors, when the clerk started calling out names. "Juror Sixty-Four, come on up, have a seat." That person would sit in the box as Juror Number One. One by one, they filled the jury seats and added alternates.

Jury selection is a tedious process in which each attorney looks for jurors likely to be sympathetic to their side—or at least neutral. If a biased or emotionally reactive juror makes it onto the panel, it can tilt the whole trial, no matter how solid the evidence is. If one side ends up with a jury that leans against them—too skeptical, too emotionally reactive, or simply not attentive—it's like starting the game down a few points. Even the best arguments can fall flat in front of the wrong audience.

Two or three rows up, sitting against the wall, Amiee noticed one potential juror, a mid-twenties Asian male, acting somewhat oddly. While everyone else was focused, eyes forward, this guy kept looking around the room. Not alarming or dangerous, just different. He didn't seem distracted or disengaged but rather fascinated, curious in a way that stood out.

Eventually, his number got called, and he became one of the twelve jurors. During selection, both sides—government and defense—get to question potential jurors. "Have you ever served on a jury? Where do you work? Have you ever had any interactions with law enforcement?" They can also dismiss a few potential jurors without giving a reason.

Meanwhile, during questioning, the Asian kid, Henry Chung, said he worked three jobs: Domino's delivery driver, call center rep for Kaiser Permanente hospitals' membership services, and a weekend security guard at Pasadena City College Library. To me, he seemed like an articulate young man, out there working hard to make ends meet and have a good life.

The prosecution kept him and dismissed a couple of others, and eventually, the jury was seated. Then came the opening statements, starting with the prosecution's outrageous version of the story. They claimed Corporal Jensen had punched a child in the face at the fair. That the poor kid had just left the petting zoo and was walking along, holding a balloon, eating cotton candy, filming his dad getting arrested because he thought it'd be funny. And then, out of nowhere, this crazed cop snapped, yanked him out of the crowd, and punched him twice in the face to "teach him a lesson" about filming police.

I was boiling inside at the blatant lies. And yet—nothing from our side. No objection. Nothing. Why wasn't my attorney jumping up and down, screaming objections? But Michael reminded me that court decorum says no objections during opening unless it's egregious. Before the trial, he had given me a pen and paper and told me to just draw tiny circles and not make eye contact with the jurors. It was hard for me to not make eye contact or smile or nod at people who were looking and staring directly at me. It felt so foreign to me because I am a very personable and outgoing person who loves engaging with people. I found myself analyzing every move and expression, thinking, *Do I look mad? Do I look sad? Do I look like a robot? Or a serial killer void of emotion?* All these thoughts just whizzed through my mind as the prosecution was up there painting me as this out-of-control, child-beating cop and overall horrible person.

Finally, it was our turn to talk about what had really gone down that night and that we were good guys, not the corrupt cops the prosecution was painting us out to be. Schwartz stood up and gave our opening statement to the jury in a very factual, easy-to-follow manner. He detailed the course of events, highlighting the facts and real evidence of the arrest rather than playing on the emotions of the jury.

Prince's attorney, Stuart Abrams, followed up with a consistent, accurate version of the story: The kid had ignored commands and walked straight through an active scene. Jensen had reached out calmly and asked him to step aside with no aggression. They even played the audio so the jurors could hear my voice, calm and controlled. Stuart keyed in on Prince's observations and actions. The swing from Alvarez to my head had been seen by him as well as the other officers on scene. He noted how controlled the officers had been and how they had done what was necessary to overcome Alvarez's resistance and take him into custody.

The next day, Juror Number Eight, Henry Chung, sent a note to the judge requesting to speak to us privately (without other jurors present in the courtroom). They brought him in alone, and the judge asked what was going on. He said that the night before, he'd been watching TV at home and had seen a news flash about the Pomona Police and the fair. Something brief—"Federal trial now underway"—just a teaser. He said he hadn't watched the segment or learned any new details but felt he should disclose it.

That kind of outside exposure is a big deal. Jurors aren't allowed to do their own research, so the judge sent the attorneys into separate rooms to discuss what to do. My attorney immediately declared, "I think we should dismiss this guy." Prince's counsel agreed. But I had a different take.

I said, "Look, this guy works three jobs. He gets treated like shit at two of them, probably all three. He's a security guard. He knows what it's like to be disrespected for wearing a uniform. That's the kind of juror I want—someone who understands the law enforcement mindset."

Michael looked at me and replied, "I work for you. I'll do what you say. But I think it's a mistake."

I said, "Keep him." Prince agreed.

What followed was a trial full of smoke and mirrors. We couldn't read the jury at all. The foreperson was a man who worked in the tech field. Another juror was a USC music professor. Another was a mechanic. One was a nurse. The rest blurred together. We thought we had a strong case that was based on facts and the evidence permitted; no way they'd come back with a guilty verdict. The trial proceeded, including closing argument theatrics and more lies by the prosecution. The jury had deliberated just one day when we learned they were hopelessly deadlocked and didn't know what to do. The judge told them to keep working at it. Next day, same result—still deadlocked.

A day later, they brought the jury in. The foreperson handed a note to the bailiff noting the eleven to one deadlock. I thought to myself, *Okay, good, there's only one juror against me. We should get through this.* The court clerk read the official entry noting the deadlock. The DOJ prosecutor asked to poll the jury to verify the guilty–not guilty balance, which everyone on my team expected to be in my favor.

It was eleven to one for guilty.

The blood drained from my body. Everyone in the courtroom was stunned. Even the clerk's eyes were wide—no one could believe it. And then came another shocker: The one holdout, the one vote for not guilty . . . was Henry Chung.

According to the jury foreperson who spoke to our attorney, Chung had refused to budge despite pressure from the rest of the jury. After a mistrial was declared and the trial ended, the judge reminded us that our pretrial confinement still applied. Our attorneys talked to some of the jurors outside the courtroom while the DOJ and FBI praised the jurors for all their great work. The USC professor even told the DOJ, "That guy just wouldn't change his mind. We couldn't deal with him anymore."

We milled about for a couple of minutes, still in a state of disbelief and denial and fear. Nicole Pifari, who was our co-counsel with Schwartz, was with us. Nicole was solid. She was a former state trooper from Montana who'd left law enforcement to defend deputies and officers from this very type of malicious prosecution by overzealous prosecutors, both state and federal. While we stood around, she listened to the posturing banter from the prosecution and shared a quick story to lighten the mood. I always laughed inside at the way the attorneys acted friendly toward each other outside of the jury. As we walked away from the prosecution, Nicole said, "I wanted to reach across and strangle that guy; he's so full of shit." The guy in question was a federal prosecutor who dressed just like Teddy Ruxpin, the little electronic talking bear-like toy from the mid-eighties that wore a T-shirt and vest and told exciting stories. The prosecutor wore a tidy gray suit over a maroon vest with a matching tie and carried an appropriately swanky leather briefcase. Nicole thought he looked like Teddy, referring to him as such everywhere except in the courtroom. "I had to sit next to Teddy Ruxpin in the cafeteria today," or "Teddy Ruxpin was behind me in line checking in at the hotel." I appreciated her sense of humor as much as her dedicated hard work.

Out in the courtroom lobby, we heard a side door open loudly, where, according to my wife, a small-statured male quickly zipped through the crowd, bypassed the elevators, and disappeared through the stairway door.

Amiee watched this all unfold, and he was gone before she realized who she had seen, exclaiming, "Hey, that was the juror. That was Henry Chung."

Henry fled down seven flights of stairs and vanished. When our attorneys met later, they learned that Henry was pissed off because the rest of the jury wouldn't stop hounding him to change his mind. He couldn't believe that the jury was falling for this ridiculous story that the DOJ was presenting.

We tried to find Chung later to thank him. We called Domino's—but no one by that name had ever worked there. I searched all over the place online, Amiee checked social media (what kid in his twenties didn't have a social media presence?), and we even checked with the college. No record of a Henry Chung working at the library. It was like he didn't exist.

I told Michael, "I can't find this guy."

And he said, "I don't think he's real. I think he was your guardian angel. The Lord wasn't ready for you to be in prison."

Amiee shared a similar sentiment: "That kid was different. Like he was seeing the world for the first time."

I believed them. No one has seen or heard from Henry Chung since.

He was a ghost. Or maybe . . . something more.

Dad's work semitruck, 1976

Me and Bill, Rio Hondo Police Academy Graduation, 1997

New cop patrol ID, 1998

Me in the motors unit, 2004

Me with my police vehicle on shift, 2006

Pomona SWAT team plaque, 2014

Me working a recruitment event, 2016

My horse, Chevy, at dusk, 2016

Chevy at dusk with our dog, Jenny

News still from the altercation at the fair, 2016

Different perspective of altercation at the fair, 2016

News photo of me, Prince, and Mike, 2017

Riding Chevy by the house, 2017

Cops 4 kids, 2017

Riding with my wife, Amiee, 2018

Federal trial legal team, 2019

My son's academy graduation, 2020

My wife, Amiee, and me, 2020

PORAC article on exoneration

PPD body cam footage of suspect with a sword, 2020

CHAPTER SEVEN

WHAT WE LEARNED

The Baton Guy and David vs. Goliath

We got a drilling during the first trial. It felt like getting your butt whooped good in a scrap with the school bully, and this fight's scars lingered for months. The grand jury process is supposed to include certain checks and balances, one of which is hearing directly from the people involved. That includes law enforcement, witnesses, even the alleged victim. Why? Because the grand jury has the responsibility of judging credibility and believability firsthand. That's the entire point. But that didn't happen here.

We found out that Agent Raye had walked into that grand jury room and spun a story full of exaggerations and outright lies—fabricated distances, skewed timelines, and critical omissions. Most importantly, she concealed exculpatory testimony from two key witnesses who had been present alongside Alvarez that day at the fair during the arrest. They were interviewed by the FBI, and they both said Carlos Alvarez had been interfering with the arrest of his father and uncle. That part conveniently never made it into her presentation to the grand jury. She went on like that for two and a half hours. Approximately an hour later, a grand jury indictment came down. It was a textbook example of fraud—an egregious abuse of power, malicious prosecution, and manipulation of the grand jury system of checks and balances, which includes commentary directly from witnesses, including police officers, fair patrons, and the "victim." The grand jury then assesses potential jurors' credibility (are they a reliable witness or full of shit?) and believability (do they use direct and confident statements or say "um" and "I don't know"?) of their testimony. This builds the foundation on which the grand jury decides to issue an indictment or not.

The FBI and DOJ steamrolled the process, plain and simple. They didn't let the evidence lead them to a suspect—they picked their target first and built a case around him. They sliced testimony, cropped video footage, and cherry-picked details until they could spin a cohesive narrative. Agent Raye didn't mention that her original distance estimate of Alvarez to the backs of the arresting officers—somewhere around ninety feet—had been sent to Quantico for verification and came back stamped "physically impossible." She just left that part out. Why correct something that could jeopardize her getting the indictment?

It was a bombshell when we found out. But it didn't stop there. One of the first major obstacles in the trial was the discovery dump—thirty thousand–plus pages handed over just days before court. Any competent prosecutor would've produced those documents in a timely fashion. Any fair judge would've paused and said, "Hold on, this isn't right, you can't dump that much material on the defense two days before trial." But that didn't happen. No objections, no delays—just "see you Monday."

Even worse, there were key FBI documents—witness interview forms from that night at the fair—hidden in that stack of pages. These documents were ultimately brought to light on the second day of the second trial, despite us having spent months preparing for both of the trials. In fact, as we were preparing for the second trial, we still had no knowledge of the interviews of the cousin and brother who had been witnesses at the scene, and more importantly, we had no idea that those documents even existed. There was so much evidence that never even made it to the first trial or was purposefully hidden, and all those lies were working against us.

Before the first trial, we'd thought, *There's nothing here. They're grasping at straws. This is just a formality.* But then the jury came back eleven to one for guilty, and everything changed. The pressure was on, and the stakes skyrocketed.

Another critical blow came when the judge barred our use of force expert. This expert was ready to testify on police training, defensive tactics, reasonable force—exactly the kind of information jurors need to understand what's normal or justified in law enforcement. The average juror has no idea and is ignorant of police work. They need to be educated as to the nature of police and suspect interactions and what we as police officers can and cannot do. They've likely never tackled a volatile troublemaker in the dark with a crowd screaming around them. They've only seen cop shows or old Western movies where the good guy shoots the gun out of the suspect's hand at the last moment or whizzes a bullet past a cattle rustler's cowboy hat just close enough to knock it off.

But the judge barred any and all testimony related to police procedure, tactics, crowd control, or the basic physics in play when fighting with or attempting to gain control of someone. Our expert was muzzled from all of that, and he was even stopped from reiterating state and federal Fourth Amendment rules regarding search, seizure, reasonable force, and internal affairs investigations. He was reduced to saying only that he'd reviewed the case and couldn't comment further. The prosecution, meanwhile, was free to speculate. "Did he grab him? Did he pull away? Did he swing?" All without context. And none of our witnesses were allowed to explain what would be considered reasonable under California law, under federal law—under any policy, really. We weren't even allowed to explain the basic rules of when an officer is allowed to take someone into custody or crowd control tactics. To say this handicapped us and set the odds against us is an understatement. It undermined the factual basis of evidence, turning the foundation of the trial into emotions rather than facts based on the reasonable officer standards.

The court's stance? Complete lunacy. They said the jury could just watch the video and figure it out themselves. As if anyone watching shaky footage from a 2000s-era flip phone with a fisheye lens could fully grasp what had happened. How could a juror make an intelligent decision, influencing my *freedom*, without having *all* the information? My attorneys attempted to state their views on fair and just proceedings, but the judge would have none of it. "I've made my ruling, and I'm done arguing about it." What could we do? What the judge says, goes.

So we pivoted. We went on the offensive, locating a video expert in Beverly Hills to scientifically analyze the fair video, breaking the video down frame by frame for anything to help my case. We found two distinct frames—just two: In one, you could see the moment I'd braced for impact from an incoming punch, my left hand raising in defense and my eyes squinting in a reaction of "oh shit, I'm about to get hit." The other frame showed me turning away, hands pushing Alvarez off, just as his fist had come down on top of me. That had never been visible before. The video's low frame rate—thirty frames per second—had hidden it all, but slowed down, it was right there, in vivid, living color.

Still, none of this mattered unless we tackled the root of the issue: the FBI and their investigation. My attorney was a "go at 'em, gloves off" type at the state trial, but it's a different game in federal court. If you're going to question an FBI agent or the Department of Justice, you better have your ducks in a very straight row.

That meant digging into the evidence trail. The so-called "Barbie Squad" had gone out to the fairgrounds with their Starbucks lattes and their power suits, measuring distances in the daylight, long after the incident. The agents were guessing how far Carlos had been from the officers' backs. The farther away they could make it, the less threatening Carlos would appear to the jury. They used the landmarks they could see in the video—the green-and-white fence, the pizza stand, etc.—to make their measurements. According to the notes we obtained, her first measurement had been over one hundred sixty feet! After some discussion with her partner, Agent Raye had sent a ninety-foot measurement to the FBI lab in Quantico to get diagrams made. Quantico had come back with a request for a better measurement. So they'd gone back out to the Fairplex and obtained another measurement—fifty to sixty feet. Again, the FBI lab had sent back another correction request. In all, they'd taken three distinct measurements and sent each to Quantico. Over one hundred feet, then ninety, then sixty or fifty. Quantico had rejected them all, saying they were physically impossible because of fixed landmarks. Only at around thirty-two feet did the math finally work. Our own expert calculated twenty-eight to thirty feet based on those same landmarks.

But context matters. The FBI measurements were done in a sterile environment, in daylight, with plenty of time to think. That was not the reality of the situation I'd faced. The incident had happened at night, under neon lights, with chaos all around. Your sense of distance isn't the same in a high-stress, split-second decision in the dark. You also have to consider the human factor, like Captain Sullenberger landing his plane on the Hudson River. In hindsight, people said he could've made it back to the airport, could've done this or that, or should've tried that other thing. Easy to say, sitting safely behind a desk. Go out among the elements and make a real-time decision under pressure, without knowing what's going to happen next, then see what you think.

That was exactly what had happened here. I saw Alvarez accelerating toward the officers from behind. I couldn't stop and check the distance; I couldn't look back. My last visual was him charging forward, and I knew if I didn't intervene, he'd get to those officers. And he wasn't going there to politely plead or ask questions—he likely planned on throwing a punch at them or otherwise interfering. That's what the history of our line of work has taught us. You don't let someone agitated and uncooperative approach your fellow officers from behind—you stop them. When I made contact with Alvarez, there was no one left between him and the backs of the arresting officers.

And yet every decision, every move, is second-guessed. Twenty-five or thirty years ago, this culture of blaming and attacking police officers wasn't there. People seemed to understand that sometimes you had to act fast and in imperfect conditions.

As for Alvarez, he wasn't some helpless kid; he knew exactly what he was doing. He said it himself: "I'm a minor." His dad even said, "You can't touch him. He's a minor." As if that made him untouchable. But if you act like a man—swing like a man—you get treated like one. The Feds knew that, but they'd still pushed their "poor child at the fair" narrative. It's insulting, but it's what they do. Craft the story first. Shape the evidence later.

Where's the Baton Guy?

Prepping for the second trial had just as many lurking legal landmines as the first. For starters, the Feds had indicted only Prince and me, while "the baton officer" was nowhere to be found in court. With three officers involved, there should have been three indictments, especially for the one seen using a baton in the video. This third musketeer was suspiciously absent, and in such a high-profile federal case, why wasn't he being held accountable for cracking down on the suspect, as seen in the video?

The short answer? Because it didn't help their case.

In police procedural terms, a baton is considered a weapon, which, if used, escalates the incident into an official "use of force" status, and that changes the game in court. When I heard they hadn't indicted the other officer, I couldn't believe it. The only reason that Alvarez had had to go to the hospital that night was because a baton was on scene; he didn't go there for treatment for any serious injuries. There were none, and he didn't claim that there were. The only marks on him had been skinned-up knees and elbows (from squirming around on the ground on all fours), but the use of a baton automatically triggered a hospital visit and medical clearance.

If the DOJ had indicted Officer Three, they would have been required to allow a use of force expert (upon which we had based much of our defense, before the judge dismissed it). Since the public wouldn't know the intricacies of what an officer is permitted to use in force, the expert would take the time to explain it in simple terms so the jury members could make an informed decision about my freedom. That was a problem for the prosecution, because then the jury would be educated on what a baton is, what it does, and what's considered acceptable use. And they didn't want an educated jury. They wanted the average juror with no law enforcement experience to see a baton and assume brutality.

It was very telling, then, that the DOJ had not elected to indict Officer Three. They'd known they would have had to answer basic questions about him as a witness—most alarmingly, why he had not filed a report about his involvement that night. But even that wasn't the real reason—they simply didn't want a defense expert in court to lay a foundation of reasonableness of force with the jury. A jury that understands the ins and outs of use of force is more likely to weaken the prosecution's case.

So Prince and I had had our legs cut out from underneath us; our team was unable to properly educate the jury and inform them about the basics of this stuff. Without someone explaining it, the jury would never know that there had been a landmark court ruling six decades ago stating that you can only hold an officer responsible or liable for what they knew at the time, not for anything based on events had transpired since then:

The United States Supreme Court said in the Graham v. Conner decision, "The reasonableness of a particular use of force must be judged from the perspective of a reasonable officer on the scene, rather than with the 20/20 vision of hindsight."

Think about this type of perception like the "bee on the shoulder" example. Two of us are talking face-to-face and a bee lands on my shoulder. If you suddenly swat it off, my natural reaction—as a uniformed officer who doesn't see the bee—is to react defensively. My body responds to what it *perceives* as a threat. That's what we're judged on—split-second decisions made with the information we have in the moment. If this is caught on camera, it will look like you are an aggressor, bringing a large, swiping movement onto my shoulder to remove the bee. From my perspective, and in the heat of the moment, the first thing I'm going to do is push you back and say, "Hey, what are you doing?" In reality, you're being a nice guy, looking out for my well-being, and after you explain what you did, I will probably thank you and apologize for pushing you back. It just goes to show that perspective is everything, and in federal court, the stakes are life in prison if you can't get the jury to understand your side.

But during a trial, they don't care about perception; it's all about a narrative. They only care about what's in the video. My memory of the incident, my understanding at the time? Tossed aside. In the state trial, Gaines, the prosecuting attorney, had played on this and asked, "Well, Officer, do you see where you threw that punch in the video?" I *thought* he'd thrown a punch at me. Later, when I saw the video, I realized it had been more like a flailing arm, maybe even the opposite hand than I'd thought. That didn't mean I'd lied—it meant I'd testified to what I'd seen at the time. Suddenly, we were not talking about my memory or my perception of the situation at the time; all the context, all the nuance—gone. Redacted. The jury only heard the word "yes." That was not justice.

This kind of thing happens in everyday conversation, of course, where our memories of the past can be incorrect. If you're reminiscing with your high school buddy about his old car, your recollection may be that he used to drive a bright red 1966 Mustang. But your friend might remind you that it was a blue 1968. Your perception drives your memory. Now, imagine that same casual conversation played out in a court setting. A prosecutor will remind everyone that sixty seconds prior, you testified *under oath* that the car was red, but after some outside information from your friend jogged your memory, you conceded that it was blue. Were you lying then, or are you lying now? To a jury, under the guidance of a skilled prosecutor, you will come across as a dishonest defendant.

There had been questions to the effect of this during the state trial:

Prosecution: Regardless of your perceptions that evening, now seeing all this on video, is it your opinion that he tried to punch you?

Jensen: Well, no. I mean, reasonably, it looks like a flailing arm. Hell, I thought it was his right hand, but now looking at it, it was his left hand.

Prosecution: So, sitting here today and having seen the video, are you willing to admit now that he never tried to punch you?

The problem with removing the heat-of-the-moment perceptions and boiling things down to a "yes" or "no" answer about a videotape means that if you do agree to such questions, they redact your previous, human-in-crisis observations and explanations and strike the testimony altogether. Then the jury ends up with a distilled and curated version of the event, and your affirmative answer to what the video shows is now front and center in their minds.

My attorney told me the only way my side of the story would ever be heard was if I took the stand myself, which goes against everything the Fifth Amendment stands for. But I had no choice, so I trained for it—three to four days a week, eight hours a day, while Michael grilled me like a prosecutor.

It was in some ways a David vs. Goliath story. Gaines had prosecuted dozens, if not hundreds, of cases involving Pomona police officers. His ultimate goal was to get a consent decree from the FBI—a feather in any prosecutor's cap—against the Pomona PD. He was so adept and cunning at taking on cops in court that I'd heard several backroom stories among the local rumor mill that he had a "black book" of inside information about officers, a report card of sorts on their courtroom demeanor: supposed strengths and weaknesses, who tended to "not recall" and who was a straight shooter on the stand. If there was a book like this somewhere, we never saw it, but it was the kind of thing Gaines was known for. He was highly knowledgeable and savvy and knew all the ins and outs of placing police officers on the stand. While he was an intimidating force that way, he was also cocky, which turned out to be his Achilles heel. Because of all this, we practiced going over mock testimony for weeks; I needed to be ready for a skewering by the prosecution.

With a pit bull like the DOJ charging hard for my throat in court, I expected to have some backup and support from my department, especially from my former deputy chief, who was now Chief Oletti. I was stunned that he had not come out before the first federal trial, as he'd said he would defend us "seven days out of seven," with the full support of the department—but I guess that support was a fairy tale. He never showed up to anything in any form, written or in person—a complete coward. When the Feds had started sniffing around the department and questioning its leadership, the new chief had been at the FBI's behest to defend himself and the decisions he made in signing off on and reviewing all uses of force. He'd found himself doing all he could to accommodate the FBI, fearing they would continue to rummage through department files and policies. That meant completely abandoning me.

The former chief and deputy chief had both reviewed the admin insight and determined that things were within policy the night of the incident. Things had all been good and well until I was federally indicted and the deputy chief took over. Unlike the former chief, Oletti was the type of man who cared more about the police chief label on his social media profile than he did about standing up for his people. I had supported Oletti throughout his whole career, but he seemed to change when he was promoted to chief. Top brass decision power, narcissism, and self-aggrandizement are not a healthy mix, especially under federal scrutiny. Bottom line: In my opinion, my new boss had a secret handshake with the Feds, so I was on my own in that courtroom and at work. That ultimately severely affected any chances I had of advancement.

Remember, it wasn't just me going through all this. Prince had also been indicted and sat next to me the whole time, each of us flanked by our legal teams as co-defendants. Standard practice in that courtroom was for the prosecution to give their opening statement, followed by my attorney's opening statement, and then by Prince's attorney and his statements. It quickly became overwhelming and repetitive for the jury, so we had to be smart about not asking the same questions.

While the drama inside the court was ongoing and eventful, we also garnered a minimal amount of media attention, with our faces and the arrest video appearing together on Channel 4 and other outlets that were trying to sensationalize the story and get headlines.

On multiple occasions, the media tried to exploit the trial and the video of the incident to garner public outrage. It never caught traction or fire, no matter how many times it ran. One reporter was out at Starbucks in Pomona polling people after showing them the video; most of their responses were something like, "The guy should've listened to the cops, and what's the whole story?" No outrage, not a single complaint from a Pomona citizen was ever received by city hall or the police department regarding the arrest at the fair.

In many ways, I miss the old guard of the LAPD. Chief Daryl Gates used to say, "If you're gonna make an omelet, you have to break a couple of eggs." Law enforcement is not a pretty business. When you have a combative suspect or someone you have to take to jail, there's no delicate way to do it. It's never going to look good on video. But sometimes, you have to use force. My generation was the last of the hard chargers, the last to actually be real cops on the street, whereas now, everything is just robotic and political.

The first trial ended in October. I spent Thanksgiving and Christmas in a fog, knowing the second trial started in January. That was a very somber holiday season because as much as I had faith that the Lord would take care of me and that I was going to be okay, I did not have faith in mankind. And I sure as hell didn't have faith in the federal system. I spent a lot of time alone with our animals and even more time in prayer, leaning into my faith harder than I had in thirty years. That was where I found some kind of peace.

But my attorney made one thing very clear: After that first trial, it was obvious from the transcripts that the judge did *not* like us—or our attorneys. He was basically a third member of the prosecution. We walked into his courtroom already at a disadvantage.

This system is broken. The FBI can bring you in, question you without a lawyer, forbid you from recording or having a witness. How is that okay in the twenty-first-century United States? That is categorically wrong. You should *never* be denied legal representation—*ever*. But they get away with it because nobody holds them accountable. That's disgusting. And unless you've got a million-dollar defense, you're done for. Think about this: How many people in the history of the feds, dating back to J. Edgar Hoover, have been convicted of ridiculous crimes or taken plea deals because they simply did not have the financial backing to adequately take it to trial? Had I not had legal defense insurance resources, I'd be in federal prison right now. No question.

And when someone like Neaderbaomer sues the DOJ for malicious prosecution? The attorney general says, "Nothing to see here." They stick to one narrative, peddle it over and over, and never break out of their own echo chamber. That's the sad part—because it destroys real people. Just like with Derek Chauvin. Do I think he *meant* to kill George Floyd? Absolutely not. Do I think drugs and Floyd's physical state played a role? Yes. Do I feel for the man who died? Of course, any death is unfortunate, but the real travesty is the media spinning the narrative to fit their desire for ratings. Did it appear that Chauvin made mistakes? Probably, but I don't think his intention was to kill someone. Remember, the video doesn't show everything, and we can't judge from the living room sofa without all the pertinent facts that the media will never give you.

Interestingly, at the height of the anti-police sentiment, the person who helped people see this from a different angle was Terry Crews, former NFL pro and host of *America's Got Talent*. He did police training with a local department. During this training, they gave him a Simunition training gun and ran him through a scenario simulation. In the scenario, he got a call about a trespasser at a truck yard. As soon as he walked up, the suspect charged at him, cussing him out (no weapon shown). Terry backed up, panicking, and ended up shooting. The evaluators asked why he'd shot the charging man, and he replied that he'd thought the suspect was going to hurt or kill him. "I had no time to think, just react."

After that, Terry became a big supporter of law enforcement because he'd *experienced* it. He'd seen how fast things happen and how little time there is to think. He'd realized people were holding officers to some Hollywood standard that isn't real.

That's why I believe we need law enforcement education in high schools. Call it "Law Enforcement 101" and teach kids about the Bill of Rights. Teach them what those rights *actually mean* when you're on the street. Go over how traffic stops work. Show them ODMP. org—the Officer Down Memorial Page—so they understand how many cops have died in the line of duty. And most importantly, put them through some realistic training.

Let them walk in our shoes—if only for a moment.

CHAPTER EIGHT

THE SECOND TRIAL

Burden of Proof and a Five-Dollar Airplane

"The court instructed you about burden of proof at the beginning of the case. The government gets to go first, and the government gets to give you their narrative, their take on the evidence. The burden of proof means this is an uneven playing field. We are going to get into what we expect the evidence to be. That's what an opening statement is.

What the government just told you is not all the evidence, and it won't be all the evidence in this case. And you are tasked with the hardest job in this courtroom: keeping an open mind. Even if we're hearing what should be a compelling narrative—because we shouldn't be here if it's not—that should be a starting point. That's how the system works.

If that wasn't a compelling narrative, if it wasn't a reasonable case after describing what they feel the evidence will be, we shouldn't be here in the first place. But that won't be all of the evidence, and it's an uneven playing field; they have the burden of proof beyond a reasonable doubt."

—Excerpt from Michael Schwartz's opening statements, United States of America v. Chad Kenneth Jensen, January 16, 2019

"Chad Jensen, you're going to jail!"

I didn't know who they were, where they'd come from, or how they knew when I'd be walking into the courthouse, but a small crowd had gathered on the steps outside for the opportunity to lob taunts at my family and me. It was just my wife, dad, and stepmom this time, so for a quick moment, I felt outnumbered, but I had a far more insistent weight on my shoulders.

I didn't know if it was a harbinger of anything in particular or simply happenstance, but during the first trial, our entire family had stayed in a big, four-bedroom suite, while this time, it was just Amiee and me, and we were crashing in a bare-bones hostel a half-block from the courthouse. One bedroom and a toilet for your troubles, sir. Welcome, and enjoy your stay.

Sparse digs aside, jury selection at least proceeded more smoothly this time around. There were a few preemptory dismissals, and then we really turned the screws in juror interviews, as it is absolutely true that your case can be won or lost in jury selection, just like Schwartz had told me during the first trial.

Both sides sifted through the jury pool, trying to find the most advantageous jurors for their side of the case. Some jurors were easy to deal with and appeared to be middle of the road; some were not as easy to understand, whether that meant their speaking, personal viewpoints, or particular skill set in the world. It was very scary to sit there and know that this soon-to-be group of twelve people would be deciding your freedom. Some of the dismissals from the jury pool by the prosecution were particularly difficult to accept, as these jurors had obviously presented themselves to the prosecutor as being far too conservative or common-sense-based to listen to their bullshit narrative.

In some cases, of course, they had clear justification, like the middle-aged female sitting in the top row, far right. When my attorney asked if she had ever been to the LA County Fair, she said, "Are you kidding? Yes, I've been there, and I'll never go back. It's full of gang members and fights and violence. The Pomona police, though, were always very professional, and they were very nice."

Objection! Objection! Objection! The courtroom erupted into chaos as the prosecution lost their minds, literally leaping from their chairs trying to get her to stop talking. The judge stepped in and called a sidebar, and a few minutes later, she was dismissed under one of their preemptory strikes. With that brief spark in the proceedings out of the way, the judge turned back to the remaining jurors and essentially said, "Ladies and gentlemen, please disregard everything that woman just said."

Which, of course, was complete bullshit. The jurors had heard it and wouldn't just "forget it." My co-counsel, Nicole Pifari, sat to my right. I squeezed her arm and whispered, "This isn't fair." (We learned later that the judge's overzealous "disregard everything" comment would come back to haunt him, altering the trial's outcome in the process.)

That moment was one of the rare flashes of brutal honesty in the whole trial. Mine was a classic case of the adage that your case can be won or lost in jury selection.

All the while, Nicole had her phone below the table, firing off texts to her assistant to track down whatever information they could find on potential jurors, things I'd never even thought of. She poked at her phone's keypad like a twelve-year-old girl texting in history class and, in the end, we had our jurors: a stoic, middle-aged man from Palmdale; a former schoolteacher who'd retired to become an over-the-road trucker with her husband; a couple of female nurses; a younger woman from East LA; and a variety of others. The attorneys for both my side and Prince's felt good about the selection, and the judge approved everything, so that was it for the day.

The next morning started straight away with some futile and flat-out ignored pleadings from our attorneys regarding the redacted testimony being admitted by the prosecution, and then it was on to opening statements. Michael reminded me again about my demeanor:

"The prosecution is going to play the same game because they almost got it last time. Control your emotions, don't make eye contact with jurors, don't show any emotions, don't shake your head. Just be cool and calm." None of that is easy to do when you're nervous as hell, scared, confused, and generally pissed off. It didn't help when our entire team of attorneys once again locked horns with the judge over Federal Rule of Evidence 106, the Rule of Completeness. This rule states, "If a party introduces all or part of a writing or recorded statement, an adverse party may require the introduction, at that time, of any other part."[2] In this instance, as the prosecution introduced a redacted version of an official state transcript to prejudice the jury, just like they had during the first trial, the defense should have had the same opportunity to bring in the rest of the report. Again, infuriatingly, the judge said, "I've made my ruling. If you have anything further to add, put it in writing."

On top of that, the judge disallowed, like in the first trial, our defense expert from covering critical testimony on police training and crowd control (the twenty-one-foot rule) so the layperson juror could understand perception and reaction and what had been in play at the fair that night. The twenty-one-foot rule is a critical law enforcement training principle developed in the 1980s by John Tueller, a former training lieutenant with the Salt Lake City Police Department[3]. The rule establishes that an armed suspect can cover twenty-one feet in approximately 1.5 to 1.75 seconds—the time it takes an officer to perceive the threat, react, unholster their weapon, and defend themselves. In theory, you are in a "zone of safety" at twenty-one feet or better. (The rule has since been interpreted from alternate perspectives and the safe distance argued to be greater.)

Nevertheless, the judge would not budge and made no bones about it. Even casual observation showed he favored the prosecution and likely had made his decision the moment he'd watched the incident video in his chambers, long before the trial began.

Sure enough, the same prosecuting attorney opened true to script with thirty minutes of fantastical tales of the innocent "child" enjoying a day at the fair with his family, when all of a sudden, a big, angry police officer, Corporal Jensen, had pulled him aside to teach him a lesson about trying to film the cops. She heaped on more lies with allegations that I'd later enlisted Prince to "help me cover it up."

I wanted to explode. But Michael countered with a powerful opener of his own, calling out a playing field far from level and giving a statement that I believe strongly resonated with the jury: "As my client sits here, he is 100 percent innocent. So if you believe otherwise, you have to go 180 degrees from innocent, past ninety, all the way to guilty. You can't stop at seventy-five or ninety or 120 degrees because anything from 179 degrees to zero degrees is not guilty."

About this time, the prosecution's star civil rights attorney, "Teddy Ruxpin," unveiled a disgusting performance on the mistreatment of the "child" at the fair, painting the Pomona PD as "gang members in blue" and even fake crying to add insult to injury. Never mind that he was pulling on your heartstrings, ladies and gentlemen, instead of gathering facts. It was all about emotional manipulation waged against the jury. I felt like I was about to blow my top, but our lunch break came along just in time, giving me a little time to breathe and take in some invaluable positive advice from my team.

I needed that mental strength because the first witness after the break was the person who'd brought us all together, the alleged victim—Carlos Alvarez himself. He got up from his seat in the back row of the gallery, walked (think arrogant strut) toward the front of the courtroom, and, as he passed between the prosecution table and the jury, gave a thumbs-up to an FBI agent in the front row—who returned the gesture. Nicole's eagle eyes caught that exchange, also noticing that the young Hispanic female juror had seen it as well and rolled her eyes in response. Once on the stand, Alvarez was sworn in but continued to engage in nonverbal interaction with the FBI agent, who made hand gestures toward him, coaching him to sit up straight or whatever else she thought would make him a sympathetic figure. Nicole immediately objected: "Your Honor, you can't coach a witness from the gallery. Can you please address whoever that is in the front row? You cannot coach a witness on the stand."

The judge finally did the right thing here, reprimanding the agent and threatening to remove her from the courtroom if it happened again.

The prosecution began questioning Alvarez, leading him down a rosy garden path, if you will, but something was different this time. At the first trial, he'd looked sheepish and nervous, sniffling now and then, playing the innocent, frightened victim. On the stand this day, he acted cocky, with his chin up and out a little, and a chip on his shoulder. Based on the eleven to one jury decision, he thought he could say and do whatever he wanted. Bad idea for him but good for us; the jury saw his real character. Everything he said was in direct contrast to what had really happened that night, and their side even caught themselves in their own seedy story when "Teddy" started another line of questioning to Carlos:

"Did you have a good lunch?"

"Yeah, yeah, it was good," said Carlos.

"Well, you and I have talked before."

"Yeah, Teddy, we've talked before," Carlos replied.

You could almost hear the entire courtroom skip a beat. The arrogant little prick's swagger was ridiculous. The over-the-top victim mentality and story they were telling was hard to stomach, and I even think the jury was getting a little nauseous of their seemingly buddy-buddy relationship. Nevertheless, this went on for a while longer, and the prosecutor ended the questioning. The judge then called a recess to give everyone a break.

I learned later that during the break, Schwartz had coordinated with Prince's attorney on a strategy. They saw an opportunity to gain some ground with this witness. They agreed to play on Alvarez's ego and arrogance, which proved to turn the balance heavier in our favor.

Schwartz began the cross-examination and took center stage with some "casual conversation."

"I heard you played some football in high school, lifted some weights?"

"Oh yeah, I sure did. I was in good shape," said Alvarez. "I did jujitsu, too, man, so I know how to take a punch." He continued to brag about trying to copy Floyd Mayweather's boxing technique and his exceptional grappling style. He related that having good balance and footwork, as well as knowing how to control leverage and absorb punches during a fight, were important. My attorney continued to work on showing the difference in who Alvarez really was—more of a Jekyll and Hyde than a defenseless young kid.

These arrogant statements by Alvarez were all on full display for the jury and in stark contrast to the person the prosecution had presented as a "poor young child" at the fair.

Abrams, Prince's attorney, began his cross-examination of Alvarez. He asked about Alvarez's high school football experience, to which Alvarez responded, "I was on the defensive line—noseguard, the biggest one on the line when I started at San Dimas High School." He went on to state that he'd been the strongest one on the team at Bonita High School when he moved to running back and wide receiver. He described himself as someone strong and fast who couldn't be caught by the bigger guys. It was almost comical how he would light up when the attorney asked him to describe himself, taking on a machismo style of speaking, laced with arrogance and an occasional foul word. One of the stories was that he could bench three hundred pounds at the time of the fair incident because it'd been football season and he'd been really lean.

Alvarez bragged about a few other "manly" things as Abrams continued his questioning, further exposing his inconsistencies and lies. Abrams referenced prior testimony and reports with such questions as, "Do you recall saying the cops 'hit like little fucking bitches' that night?" He admitted to saying it, thus illustrating his dishonesty on the stand with the prosecution and obviously misrepresenting himself.

Abrams masterfully set the hook in him, and I remember walking out with my family at the end of the day with just a bit more hope.

The next day, the prosecution called Dr. Taylor, who'd allegedly treated Alvarez the day after the arrest and at two follow-up appointments, one six months after the incident and another a year after that. These doctor exams had not consisted of any actual examination for a mysterious lip injury, just self-reporting. Our attorneys completely exposed him as unreliable—he had performed no actual examination and had no medical report from the date in question.

"You're pulling all this from memory from four years ago?" Schwartz asked. The doctor had no concrete answer and looked foolish, which again worked in our favor.

"You have no medical report of an examination generated for the date in question here. Did you see him on that day? How do we know we even saw him on that day?"

The part of the alleged injury that infuriated me was the fact that Alvarez had made no mention at the hospital of any lip injury, and no photo or statement about this injury had been taken by the officer at the hospital the night of the arrest. Of all the photos taken by the police, none showed any injury he could've received from me. Miraculously, according to the prosecution, Alvarez had had his mom take a photo of the inside of his lip the day after the arrest, but there was no reference photo to show that it was actually his lip and not someone else's—just a picture of the inside of someone's lip. Complete garbage and further fabrication to compel and sell the narrative.

Michael made the guy look silly. This was a nice bonus for us, as I believe the prosecution was foolishly resting on their previous verdict of eleven to one.

Next on the stand came Mr. Giles, vice president of security for the LA County Fair. I'd known him for fifteen years from my work scheduling security with the police department for major events. He testified solidly about the importance of police presence at the fair, explaining how attendees feeling unsafe can devastate attendance. The prosecution's questioning dragged on with an endless slideshow about the fair's layout, every little detail of the alcove in question, and more verbal detritus. I wasn't the only one glad to see that slog of a day come to a close.

When we reached Friday morning, Rod Hobson took the stand. Mr. Hobson was the person who'd filmed part of the fair altercation. On this day, he seemed almost high during his testimony as he went about using all the prosecution's key phrases, claiming he'd known "this cop was out of control" and "going to beat this poor kid." I remember my co-counsel writing "Do you think he's high?" on the notepad in front of us. Michael ran with this during his cross.

"Mr. Hobson, do you ever smoke a little weed?"

"Oh, yeah, yeah," Hobson replied.

"The night you went to the fair, you said you had a couple of beers. Is that correct?" Michael asked.

"Yeah, yeah."

"You happen to smoke any weed before you went in that night as well?"

"Yeah, yeah, yeah. Me and my wife, man, we had a bowl right there in the car."

His credibility dropped like a rock with every question, and the prosecution quickly sat him down before he could do them any more damage. The prosecution then called the officer who'd transported Alvarez to the hospital for clearance after the baton strike. He testified that Alvarez had looked, talked, and acted like an adult—not like a child, as the prosecution had portrayed him. Next came Sergeant

Moeller, who'd been present during the incident but hadn't intervened. Unfortunately, his testimony was obtuse, and his long-winded answers weren't very helpful. After he got down from the stand, we went to lunch, and that was the first time I'd had a bad feeling in the trial thus far. I couldn't really tell how things were going—and what my life would soon look like.

That afternoon, the prosecution called up our old nemesis, David Gaines, the slimy attorney who'd gone after me in the state trial. They went through several severely redacted pages of my testimony from the juvenile state trial. It was almost comical when they would put up the redacted pages on the overhead projector. They would read Gaines's full questions from the document, but my answers had been cut down to single words, with entire paragraphs blacked out. Nothing to see here! Oh, the words behind the blacked-out lines? Yeah, don't worry about those. It was ridiculous how they could lie and manipulate the pages to produce a narrative that they liked. Imagine what the jurors must have thought: *Why can't we see the blacked-out information? What are they hiding?*

The prosecutor continued with this charade for the better part of two hours, and one thing that struck me particularly was Gaines's arrogant demeanor, always acting like he was the smartest guy in the room. When I looked at the jury, I could see some slight eyerolls and general disgust with Gaines's delivery. The prosecutor paused her questioning of Gaines and asked for a break. The judge asked her how much longer, and she related that she was wrapping up. Right before the defense cross-examination began, the judge said, "Before we start with cross-examination by the defense, let's go ahead and end the day a little early today," and sent the jury home for the weekend, allowing the prosecution's narrative to sit with the jury unchallenged. It was an intentional ploy to leave the jury with a tarnished image of the evil police officer, and it was bullshit.

But it was all about to change.

The Bombshells

On Monday morning, the judge returned with a completely different demeanor, all full of piss and vinegar, imposing his will and even objecting to our attorneys' questions himself. Then came one of the pivotal moments that changed the trial's trajectory. During questioning, the prosecutor asked Gaines if Corporal Jensen had ever tried to offer explanations during his state trial testimony. "At any time when you were questioning the defendant Jensen in the state trial, did he ever make any attempt to offer any explanations to his testimony or correct any testimony?"

"No, he did not," stated Gaines.

What the hell? That was an out-and-out lie that had all my pistons firing full steam. I almost jumped out of my chair, like something had shocked me with electricity. I looked at all the counsel at our table, thinking surely, they must have heard that, but no reactions. I scrawled a "Did you hear that?" note to my co-counsel, Nicole. She shook her head no, and I whispered to her that we had to take a break right now! She motioned to Michael, and he requested one. During this short break, I brought it up, but no one else had caught it. "Chad, I'll be honest with you. I didn't hear that question," she said. Michael didn't hear it either, nor did Prince or his attorneys.

I told Michael, "I know what I heard."

He said, "Listen, I work for you at the end of the day, and I will do what you tell me to do. But I'm telling you if you're wrong—if we go back into court and you ask for the court reporter to reread this and you're wrong—and you're questioning a bar-certified attorney, basically calling him a liar in open court, you better be right because the judge will absolutely throw the book at you."

"I know what I heard, Michael."

When court resumed, Michael asked a few random defense-related questions, then asked Gaines about the exchange with the prosecution. More specifically, Michael asked again if I had attempted to clarify any answers in the state trial. Gaines reaffirmed his answer: "No, he did not."

Michael then requested to enter Defense Exhibit 12 into evidence—my *unredacted* state trial transcript of page 89. The judge was furious but allowed it as Michael explained the evidence would essentially impeach Gaines' testimony because he'd just said something that would be borne out in the unredacted section as untrue. The bailiff handed the unredacted page 89 to the judge, and if you could ever see steam coming from somebody's ears, this was that moment. The overhead display showed a couple of paragraphs of unredacted testimony where I had clearly explained my perception of part of the altercation that night—information the prosecution had deliberately concealed from the jury. Our team attempted to have the unredacted version of page 127, my testimony describing what I'd perceived as a punch by Carlos, admitted as well, but the judge just said no once again. Michael inquired as to why the judge would not let this additional item of evidence be entered, and the judge responded, "Move on." This additional page would have provided even more clarification and additional explanations from me at the state trial and would have been a key piece for the jury to get a further understanding regarding my perception the night of the arrest.

Divine intervention. The prosecutor hadn't needed to ask that question of Gaines, but she had, and by doing so, she'd opened the door to expose their deception and covering up of the truth. We were just praying that someone on that jury was paying attention to the whole exchange and introduction of page 89. Michael helped it along with a final statement to the judge regarding him not allowing page 127: "Your Honor, this goes to impeaching this witness's testimony and further adds clarification to my client's state of mind that evening and that his experience was denied. Your Honor, basically, my client is on trial here for something that is absolutely occurring right here. They are knowingly concealing exculpatory evidence."

What do you think the judge's answer to this was again? "Move on."

After a final series of cross-examinations of the FBI agents, court was adjourned. As I walked over to say hi to my wife, Neaderbaomer stepped in front of me and said, "I need to talk to you right now."

"Okay, what's up?"

He motioned for me to wait for the FBI crew to leave, then dropped this on me: "Where are the 302s for Alvarez's brother and cousin?"

I was confused. "What are you talking about?"

"The FBI interviewed these guys. I read the reports," he said. "They're in those thirty thousand pages of discovery. Have your attorney call me later."

That night, Schwartz reviewed copies of the FBI 302 reports, which confirmed a truer representation of what had happened, all there in black and white, and completely contradicted the prosecution's narrative. If we'd had these in the first trial, it would never have ended like it did. But we had them now, and the next morning, all our attorneys were buzzing. In fact, the entire courtroom felt electric. Michael stepped over and said, "Chad, I'm so sorry we didn't see this, but you're talking about two pages inside thirty thousand. And this was two days before the trial. I had all three of my assistants going through these trying to find it, and we didn't find it. I'm so sorry."

It didn't matter—we had them now! I was really starting to feel like the enormous yoke on our shoulders was starting to lift and small rays of sunshine were beginning to illuminate the courtroom. It's amazing what a little bit of hope can do for your spirits.

Primed with new energy and confidence, we called Alvarez's cousin, Gus Jr., to the stand in a surprise move. He'd actually been subpoenaed by the prosecution and was present, but the prosecution hadn't used him. He settled into the witness seat, and we set the hook again.

"Did you talk to the FBI a couple of weeks after the incident?"

"Yes, I did."

"Do you remember saying what happened?"

"Yeah, sure, that it was a great night."

"Have you ever seen this FBI 302 document with your statement?"

"No, I don't remember being there for that interview?"

"Okay, then we'll refresh your memory a little."

In the 302, we came to find out that Gus Junior didn't necessarily like his cousin Alvarez all that much—they didn't hang out, and Alvarez was kind of a hothead. This, of course, painted a more accurate picture of what had gone down that night at the fair. Gus went on to admit that his cousin had been hostile, yelling profanities at police and attempting to interfere—completely impeaching the prosecution's star witness.

We then called Isaiah Jr., Alvarez's brother, to the stand to refresh his memory. He'd been there the night of his dad's arrest as well and had been a member of the group of agitators that slowed down and disengaged when the police showed up and started shouting commands to get back. Unlike Carlos, Isaiah had actually listened to the police. Schwartz went over the FBI 302 interview, and as you may have guessed, every answer to our line of questioning was "I don't remember." The responses that Isaiah Jr. gave to the defense's questioning further impeached the ridiculous testimony he'd given via the prosecution. There it was again, in black and white. The prosecution's lies and false narrative exposed for the jury to see. You could see that the jury was taking notes and paying closer attention when the witness got a little testy with his answers to the defense questions. Hallelujah.

Now, let me bring you up to speed on the FBI's crooked tactics. The FBI doesn't allow anyone to videotape or audiotape their interviews but rather has an agent take notes. This point alone was shocking to me. How can the FBI interview a witness to an alleged federal crime and not tape the interview, when your average street cop has to tape a simple theft suspect in any interview they conduct? Just another example of how the Feds control the narrative—no audio or video can contradict anything they write in their official reports. It's simple plausible deniability; they can spin the interview and report to fit any angle they want.

We then called FBI Agent Raye, who had written the report, to the stand. Remarkably, she claimed not to remember writing it despite her name being on it. Michael set things up:

"This is your report, correct?"

"Well, I don't remember writing it," Raye replied.

"Is this your name at the bottom?"

"Yes."

It was bizarre to see someone with the alleged credibility of an FBI agent sit there and utter bald-faced lies under oath. Michael went on to clarify that the fair incident had been her first case with her particular investigative unit. He then set out to expose the flawed FBI interview process by asking her to explain how it works.

She explained that witness interviews involve three agents: one agent asking questions, one agent taking notes, and one agent compiling the report. In this case, the notes on the witness interview clearly stated "obs interfere with arrest"—meaning Alvarez had been observed interfering with the arrest of his father and uncle. When Agent Raye was asked why the "interfering with arrest" statement had not made it into the official FBI 302 report of the interview of the witness, she just responded, "I don't know." Obviously, this information was contradictory to the FBI's narrative and had been intentionally kept out of the official report. Once again, this showed what the FBI and DOJ would do to win.

The interview notes page that we'd finally found after so much deception was gold and basically blew the prosecution's case apart. I could see why they didn't want us to have it!

Michael then moved on to distances. One of the key elements of the prosecution's case was that Alvarez had never been close enough to the original officers to create any hazard. Agent Raye testified about measuring the distance, admitting that their first measurement of ninety feet–plus had been presented to the grand jury. But FBI headquarters in Quantico had rejected it based on immovable physical barriers at the fair location. Raye and her partner had taken a second measurement of fifty to sixty feet.

"Do you know for sure if it was fifty or it was sixty?" asked Michael. "Because what you're doing is you're giving yourself the benefit of ten feet either way, but my client doesn't get the benefit of ten feet when he's making that decision within split seconds in the dark."

But the second measurement had also been rejected by Quantico. The third measurement had been thirty feet.

Another bombshell came when Michael asked: "You used the ninety-foot measurement to obtain an indictment on my client, correct?"

"Yes."

"When you got notified by Quantico that the ninety-foot measurement was incorrect, did you go back to the grand jury to correct that statement?"

"No."

Hard stop. Michael looked up to the judge and attempted to explain that that was what I was on trial for, but the judge, per usual, just said to move on. Nevertheless, the point had been made. When Raye left the stand, I felt really good about this information coming out. Once again, it truly felt like the heavens were parting and sunshine was beaming into the dark.

Another crucial witness we called to the stand was Steve Smaul, the FBI's cell phone analyst. He testified to a twenty-three-second gap between the end of Alvarez's video and the beginning of Hobson's video—exactly the moment when I'd been trying to reason with Alvarez before the physical altercation occurred. Although he was a subpoenaed prosecution witness, the prosecution hadn't used Smaul because this evidence didn't help their case. The fact that I'd spent those twenty-three seconds escorting this guy to an alcove, trying to reason with him and be nice until he'd finally pushed off and escalated everything was not what the prosecution wanted to show.

The Second Trial

However, this was exactly what we wanted to demonstrate: I'd been calm, cool, and collected, not the rogue, violence-crazed child beater the prosecution was trying to paint me as.

We then called Officer Parker Hardy, who had experience on patrol, had worked the streets of Pomona as a cop, and had worked many years at the fair. He'd been there that night, not directly involved in the physical altercation but on scene when it had happened, and observed most of it. He did an outstanding job with his testimony, testifying that "the fair is like the Wild West" where situations can spiral from mundane to disaster in seconds, requiring immediate, decisive action. Hardy was dismissed, and the judge ordered a break.

Steeled away in our back conference area of the courtroom, Michael said in a very somber tone, "Listen, this is it. The defense is going to rest. So you have to tell me right now if you want to testify. In my professional opinion, you don't need to. I think we've got enough here with what we've had this morning, and I would leave it alone. But if you want to, you can."

I had wanted to tell my story after years of others telling it for me. Would it be the right thing to do? Prince was in the same boat and said he didn't know which way to go. I told him, "Man, I think we just need to give it up to the Lord and say, 'If this is your will, it's your will.'" For a few minutes there, it felt like I was William Wallace in the climactic scene in *Braveheart*, experiencing the one chance to fight for my freedom. That rang through my head like church bells. Should I get up there and be Wallace or just let it roll?

I'm good, I thought. *Let's let it roll.*

After this, the judge announced to the jury that they had heard all the evidence, and the only thing left was closing statements. "Today, you're going to hear closing statements from the prosecution and the defense for Prince Hutchinson, and tomorrow, you'll hear the closing for Jensen."

I cannot explain the amount of stress and pressure that I felt closing in on me knowing that I would not be testifying and that the defense had rested. It was like a ton of weight had just landed on my shoulders. I couldn't help but think to myself, *Did I make the right decision?* I was looking at forty years, and forty years in federal prison is a long time. Schwartz had explained to me earlier when we'd been talking about possible outcomes of the trial and what would happen if I were found guilty. He said that in the possibility of a guilty verdict, based on the judge's demeanor toward me, I would be sentenced to the maximum time in federal supermax in Colorado. I asked him what an appeal would look like, and he responded that it would be close to eighteen to twenty-four months before I could possibly expect an appeal to go through. Mind you, all that time would be spent behind bars. That's a heavy weight.

I just hoped I'd done the right thing.

Reckoning

The prosecution attempted one last Hail Mary to pin me in their closing. They continued to harp on their lie that the "child" had been "punched twice in the face." The theatrics employed by the DOJ prosecutors were almost comical but infuriating at the same time. I couldn't believe they could stand up there telling the jury complete lies. I'd never punched anyone that night, yet they can replay that same statement time and time again. They had nothing based on fact or supported by evidence; they just kept playing the video over and over, attempting to sway the jury to a guilty verdict using emotion and dramatics.

Schwartz's closing was brilliant and very detailed—it didn't rely on manipulative tactics, and the facts brought the truth to the surface. We started with a snapshot taken from the fair video that showed me in a defensive position and Alvarez in an aggressive swinging position. This countered the prosecution's overuse of the still photo from the video that depicted the second before the first elbow check to the sternum of Alvarez. We'd saved the snapshot for closing and had not

The Second Trial

used it in the evidence portion. We wanted to have a fresh picture that would stick with the jury in deliberation.

Schwartz reminded the jury of the fraudulent and seemingly purposeful ninety-foot distance measurement provided by the FBI agents to the grand jury to gain the indictment and how they'd never corrected the misinformation, even after FBI headquarters had notified them of the wrong distances—twice! He detailed how my report, written the night of the fair incident, matched eyewitness testimony, line by line, movement by movement. He broke down the concealed FBI 302s and witness statements so that the jury could actually see that my report was factually correct.

Another point that he made to the jury was, "What is the government hiding? Why are they only providing you with redacted state trial transcripts? Why do they not want you to have the entire unredacted transcript?" It was obvious that the DOJ was cherry-picking, controlling the narrative through redactions and concealing the truth of the case. The truth is that there was no case at all. It had all been contrived through the process of "find a suspect and create a crime to fit."

Trial closing statements usually run about forty-five minutes, but Schwartz pushed past an hour before the judge stepped in and asked for an ETA, giving him up an hour and a half before he'd be cut off. My head swirled with questions no one could answer: *Was that enough time to present our case of innocence to the jury? Does the jury know what's at stake here? Do they know if I'm found guilty of falsifying a police report, it's not just a slap on the wrist—it's a forty-year prison term at 80 percent time served?*

Michael really drove home his point to the jury with three key details. "There are three reasons this case should insult your intelligence," he said, emphasizing the inconsistent distance measurements, the eyewitness testimony that matched my report, and the FBI's attempt to conceal evidence. It all seemed pretty straightforward to me, but I sat there helpless to do anything about it, unsure what the jury thought about it all.

Closing arguments done, the bailiff escorted the jury to their "back room" to deliberate, and we were left to wander around. In these situations, trial parties are not allowed to leave the courthouse in case of additional jury inquiries, so we migrated to a big lobby with huge windows looking out over downtown LA.

Sitting on the bench outside the courtroom, literally waiting for a group of people three floors above to determine the rest of my life, an old friend I used to work with walked over and said, "Dude, how are you so calm right now? You're in a tough spot, and you're waiting for the jury's decision. I don't get how you can be so composed."

I just looked at him and said, "Because there's nothing I can do about it, man. Nothing I can do. I just have to accept whatever happens. If the Lord says I'm going to prison, then I'll go. If that's where I'm going to die, that's where I'll die. As a fifty-year-old man, I know I'm not coming out of a forty-year sentence in good shape. But I'm at peace with it. We've fought the good fight, done everything we could. My wife and I have prayed. We've said the Lord's Prayer together, prayed to St. Michael, and in the end, all we can do is pray for truth to come through." As we walked out of the plaza that day, I couldn't help but think that if my days of freedom were truly numbered, this could be the last night I would spend with my wife and the last time I'd get to hug all my kids. The weight was incredible, and my ears just rang again and again: "The defense rests."

We didn't hear anything else that day, so we left to walk back to the subway and take it across town to our hotel. Amiee couldn't take the bachelor-style hostel, so we'd relocated to a more suitable location. It turned out that it was meant to be, and God was talking to us yet again. We saw the judge walking onto the subway platform in front of us; thankfully, he got on another subway to Union Station. It was weird, though, seeing him just casually walking along, not caring about the absolute misery he was fully willing to impose on me and my family.

Eventually, we got off the nasty downtown LA subway system and emerged onto the sidewalk a few blocks from our hotel. I was somber on the outside and internally petrified. My eyes had that somewhere-else glaze when they suddenly regained enough focus to spot a five-dollar bill lying on the sidewalk. But not just any five spot; this one was folded into a little paper airplane, "parked" there like it had just come in for a landing. At first, I thought, *What the hell?* But then it hit me. The number five has a very distinct meaning to me—my mom died on the fifth day of November on the fifth floor in the fifth room of a hospital in Mesa, Arizona, and May 5 was her wedding anniversary. So anytime a five showed up in my life, it had always been a little nod from my mom to me.

We picked up the paper plane, brought it to our room, and just stared at it in silence for a minute, like we were waiting for an explanation of how it had ended up on the sidewalk. I guess you could say a lighted runway of sorts showed us the meaning of it all—it was my mom telling me, "You're going to be okay. I got you. I'm here for you."

And she wasn't alone in showing me good signs. My dad and I had both done a 23andMe DNA test, his about a year prior and mine just a month ago. I shared earlier in this story that I didn't grow up in a *Leave It to Beaver* family; honestly, I never really knew if my dad was really my dad, and I wanted to know. He had taken the test as well and I got my results back that night as we sat down for dinner—the same day the five-dollar airplane landed in front of me.

I was carrying along the thought of the unopened 23andMe email notification I had just received, holding information that would answer a lot of questions about my life and bring some light into it—or not. It surprised me how jittery I was, almost more so than on the seventh floor of the courthouse, but I finally made a nervous announcement: "I'm just going to open this email and see if, truly, my dad is my dad." A mix of relief, a couple of tears, and pride welled up when I saw: Stan Jensen, father. The results confirmed my parents' identities, and coming around full circle in that way, even with looming legal drama still to come, helped me feel more sure that we were going to be okay.

The Verdict

The next morning, Amiee and I drove to a stark gray parking garage that boasted a commanding view of the courthouse. The day had dawned clear and bright, and I remember standing there for a few minutes together, uncomfortably taking in the sunshine. *If I'm found guilty, this could be my last day of freedom. I'll never see the light of day like this again.* I'm not sure there's a way to put into words what those emotions felt like, but I obviously couldn't just point the truck south and flee to Mexico. It was time to face my future, whatever that might be.

We all filed into our familiar courtroom locations, and the judge briefly took the bench to say, "We're going to finish up today by three o'clock, so chances are, if we haven't had any action by 12:30, we'll probably excuse and come back Monday." I couldn't believe how this guy continued to toy with other people's lives. He was thinking ahead to the weekend rather than the matter at hand. In a frustrated, concealed huff, I walked with Amiee back down a few floors to the concourse area, killing time while the jury deliberated. The courthouse building is constructed in a simple square of various offices and other rooms surrounding a hollow center. At one point, I looked into the void to see Schwartz walking laps two floors up, just circling and pacing in his own silent world. I couldn't imagine the pressure—he needed to knock this out of the park. I pictured him at home plate, waiting for the pitch, and hoped he could Hank Aaron this thing halfway across the Pacific. It looked like he had fallen into a trance, and watching him for a few minutes oddly brought me into a temporary calm place as well.

About mid-morning, we got a notification from the court clerk that the jury had a question for the judge regarding evidence in

The Second Trial

the case. We all responded to the courtroom, and the judge read the question to both sides. The jury's question was actually another request from the jury foreperson for the full, unredacted state trial testimony. And true to form, the judge denied it. He stated, "Minutes from the suppression hearing (state trial) are not admitted as evidence and therefore will not be provided." I was no longer surprised by his nonchalant demeanor and, by this point, didn't have the energy to gripe about it.

After the question had been answered by the judge, we returned to the foyer to continue to wait for a verdict. A few hours later, everything abruptly changed when I noticed my co-counsel on her phone nearby, visibly animated about something. She wrapped up the call and rushed over to us. "The jury has another question. Let's go." Oh, shit, now what? We hustled upstairs and saw the attorneys literally running into the courtroom. Schwartz came up behind me, and when I turned around, he looked pale, like he had just seen a ghost or something. He floored me with a simple statement, "It's a verdict."

It felt like all the blood drained from my body. I gave Amiee a hug and kiss and whispered, "This could be the last time I get to hold you." I was now one big bundle of tensed-up nerves. Michael and Nicole hugged me and tried to be uplifting as the jury shuffled in, all of them looking at the floor, making no eye contact with me or anyone else in the room. I thought, *Oh my God, this isn't good.* Everyone was seated, and the judge had begun addressing the jury when Michael leaned over to me and whispered, "I want you to know I did everything I could." You could've knocked me over with a feather.

The judge requested the verdict from the bailiff, flipped through the pages (far too quickly, it seemed to me), and threw it down on his bench. When the bailiff said, "All rise," I nearly couldn't. My noodly legs felt like they had no chance of lifting the weight of all this.

The judge announced, "Madam Clerk, go ahead and read the verdict," and everything around me seemed to turn very dark. I was flanked by my two attorneys, with Prince and his team at the next table. I'll never forget what I heard next:

"We the jury, on the above title of action as to count one, abuse under color of authority . . . (dramatic pause) . . . find the defendant, Chad Kenneth Jensen, not guilty."

Emotion absolutely took over as the clerk continued. All counts against me and Prince were returned not guilty. Interestingly, the judge stormed from the courtroom to his inner sanctum chambers. Good riddance, as far as I was concerned. I grabbed Nicole's arm and said, "Is it over? Is it really over?"

"It's over."

My face was soaked with tears, same with Prince's, my attorneys', and even the clerk of court's. I hardly knew what to do next after being in such a state of confusion, fear, and anger for so long—feeling for five years like I was locked in a room with its walls closing in on me, inch by inch, every day. Astonishingly, and to my confusion, the father of Carlos Alvarez approached me in the gallery of the courtroom as I was hugging my family and wanted to congratulate me on the verdict and give me a hug. My son stepped in between us and told him that now was not the time, that maybe outside in front of the courthouse would be better. He understood and walked out.

After a long round of hugs and backslapping congratulations all around, we were walking toward the courtroom gallery when the clerk of court met us halfway and said, "I cannot tell you how happy I am for you. This was so wrong on so many levels." I didn't have words of thanks worthy enough, for anyone, to convey how it felt to leave that courthouse a justly free man.

And I don't have the words here to do any justice to the feeling of walking out the doors into glorious sunshine. I had never felt so good.

Redemption

In an interesting plot twist, when the second trial wrapped, I was with my wife, my son, our attorneys, and Prince and his family outside on the courthouse steps waiting for the jury to come out. Carlos's father approached once again and said, "I just want you to know that it was never you, man. We never thought you did anything wrong. We wanted Dossey. It was the FBI that wanted you." Dossey was the original arresting officer of the father.

I tried to contain my inner "are you fucking kidding me" as he added even more. "Hey, would you be able to call Carlos? He really looked up to you and could use some mentorship."

I was stunned, to say the least. Couldn't believe what I'd just heard, coming from the same guy who'd taunted "Fuck you, pigs" at the fair, and finally said, "No, I'm not calling Carlos."

Ironically, a few months later, we saw Carlos's father on the news opining that his son never got justice. However, despite the contradiction, it was heartwarming in a way to receive that vindication from him. Regardless of his character, to hear that directly from him—not through someone else or a text—meant something, and I give him absolute credit for doing that.

Ten minutes later, the jurors trickled from the courthouse, and we spoke with a few of them to get more insight into what all had gone down in deliberations and how we'd made it this far. One fascinating element was that the younger Hispanic (we referred to her as the East LA juror) woman walked right by us and went down to the curb line, appeared to be waiting for an Uber. Again, not being able to read who she was or anything like that, I said to my wife, who is also Native American and Hispanic, "Hey, maybe go see if you can talk to her? I don't want to walk up to her and scare her. Would you mind just asking her what she thought?"

After a short conversation, Amiee came back and shared that our mysterious juror had said, "Oh, I had it from the beginning. If the cop tells you to stop, you stop. That kid was wrong."

Wow, you really can't judge a book by its cover. There'd been several times during the trial when she would sit there and spin around in her chair, twirl her hair, and look around the room like she wasn't really paying attention.

We were watching our silent cheerleader ride away in a chauffeured car when the jury foreperson walked up and said, "I need to talk to you. I have to tell you everything." Hell yes, we wanted to hear this.

"I'll tell you right now, as the jury foreperson, there was no way that I was going to have either you or your partner go to jail for that guy. This was absolutely crazy." And her big point was that she'd known the prosecution had to be hiding something. Come to find out, the foreperson had served on a jury prior to this one and just didn't trust the feds "because they hide things from you."

"On that first day of deliberations," she said, "I sent in the letter requesting the unredacted copy of the full state trial transcripts, and they wouldn't give it to us. So right then, I knew they were lying to us." She went on to tell us that their initial blind vote was eight to four for not guilty. After discussion, it moved to ten to two, then eleven to one, with the holdout being a young man in his twenties. The foreperson had essentially relitigated the case in that back room, asking each juror why they'd voted the way they did and explaining over and over how I had attempted to diffuse the situation at the fair and how the feds had outright lied under oath, in addition to withholding evidence. When that final holdout had seen that the judge refused to provide the unredacted testimony, he'd finally come around, putting us in the clear.

With room to breathe and hear this recount of the process, it was fascinating to learn what goes on behind closed doors that determines fates and rewrites scripts of life. But in that moment, all I knew was that mine could keep on truckin' in whatever way it deemed proper and true.

We migrated to a nearby pub for a few celebratory beers, and while walking back to the cars afterward, I jumped and ran around in my own rendition of George Bailey in *It's a Wonderful Life*, shouting, "Merry Christmas, parking garage! Merry Christmas, federal courthouse!" It got even better on the way home when I received a call from the internal affairs lieutenant. He had consulted with the head of IA for Los Angeles County to determine what to do with me now. The IA boss had advised, "Act like it never happened. These guys are exonerated. Get them back to work right away."

Hell yes. Now, my expectations started ramping way up. A few hours earlier, I'd been in the depths of despair, but the clouds finally parted to let some light in. Our judicial system is the best in the world, but it remains flawed, especially at the federal level. Cases like this are rarely won—mine nearly wasn't. That's why this story needs to be told.

The Dismissed Juror

It really is a small world. My son Nicholas was a car salesman at a local dealership around the time of the second trial. After the trial ended, the newspaper ran a little quarter-page article with a picture of us headlined "Pomona Officers Exonerated in Fair Incident, 2015," or something like that.

Nicholas had cut that article out of the paper and put it in a clear picture frame on his desk. Months after the trial, a fellow employee from the service department walked by his desk and saw the article.

"Oh, hey, do you know that person?" she asked, pointing to the picture.

"Yeah, that's my dad, Chad," Nick said.

"Oh my gosh, I didn't know that!" She went on to explain that she had been a potential juror on our case. She was the one who'd explained the LA County Fair was a scary place.

Nick said, "I want you to know you made a difference. What you said and how you said it made an absolute difference. You were one of the only people in the entire trial who accurately portrayed what the fair really is. Every other time, the prosecution made it sound like some small-town fair, and that's not what it is at all."

Tears welled up in her eyes, and she told Nick, "I felt so bad about the whole thing. After I got excused from the jury, I read all the news coverage about the case. When the trial was over and I saw he'd been exonerated, I was just so happy because the police had always been nothing but professional with me."

Of all the places she could have worked, of all the people my son could have met—there she was, right there in the same building, and neither of them had known about the connection until she saw that little newspaper clipping on his desk.

In the very beginning of this case, the government gave you a narrative. At the end of the case yesterday, [the prosecution] stood up, looked you in the face, and said, "This case is disturbing." He said that twice. He wanted to explain why the government finds this case disturbing.

It's ironic. That's probably the one thing I can agree on with the government—this case is disturbing. But not for the reasons they gave. This case should disturb you. It should insult your intelligence.

Now, the court's instructions—especially the one we call the reasonable doubt instruction—tell you that reasonable doubt is a doubt based on reason. It's not nonsense. You can have a doubt if you can articulate a good reason for it based on the evidence. But there's a second part to that instruction: It also says reason and common sense. We don't leave common sense at the door. Yet throughout this trial and the entire investigation, that's exactly what they've been asking you to do. That's why this case is disturbing.

You can say this is a close case. You can say it's a hard case. But it's not anymore. It really isn't. You should be insulted by this case.

I'm going out on a limb here—because you don't know me, you don't know my client, and you don't know anyone else in this courtroom. You've never met us before. But this government—and I say "the government," not just two people trying to present a case—they have all the resources they want. We all know that.

And with those resources, time and again, they chose not to put things in front of you. Time and again, they chose to spin what was in front of you, to plainly ignore it unless it helped their case. That should insult your intelligence. That's disturbing. That's disturbing. They have a burden of proof. We have no burden.

This case isn't about which narrative makes no sense. We don't have a burden of proof. Their narrative—their position—must make so much sense to you that it rules out any interpretation that points to innocence as reasonable. If there's a reasonable interpretation of this evidence that points to innocence, then it doesn't matter whether their narrative makes sense or not. He's not guilty.

We are going to go over the law the court read to you. We're going to go over the different elements. We're going to marry those elements to the evidence in this case. If they only meet two out of four elements on a charge, he's not guilty. They have to prove all of them.

—Excerpt from Michael Schwartz's closing statements, United States of America v. Chad Kenneth Jensen, January 24, 2019

CHAPTER NINE

EXONERATION TO ISOLATION

Furniture Movers and Combat Flights

Pomona Police Department, Monday morning, 0800, two days after my not guilty verdict in federal court. I had a spring in my step on the way to see Lieutenant Conner from internal affairs in his second-floor office. We'd spent a lot of years together coming through the ranks, and his friendship had enabled me to approach the job in a balanced way—work hard and be involved in the community, but don't brownnose your way up.

I wanted to retire as a lieutenant someday, but it was also important to me to hold true to my own character. When I sat down in his office that Monday morning—about thirty minutes before the chief was scheduled to arrive—we started by debriefing what had happened on Friday and over the weekend, particularly our phone conversation on my drive home about getting back to work.

"Before we get started today," he said, "there's an elephant in the room, and I want to get it off my chest."

"Okay, what's on your mind, brother?"

"We both know that you're the one who should be sitting in this chair. You're the one who should be the lieutenant here, not me."

His words hit me hard. He continued, "You have always been one of Chief Oletti's staunchest supporters throughout his career as he climbed the ranks—from captain to deputy chief to chief." Oletti had inevitably made enemies along the way. Everyone does when they climb that ladder. But I had always stood by his side, even to my own detriment. On more than one occasion, that loyalty had kept me from getting promoted to sergeant because I'd valued my friendships and remained loyal to the people close to me. "I just want you to know that's not lost on me. I know this."

"Hey, man, I really appreciate you having the guts to say something like that."

With that icebreaker conversation behind us, we prepared for the nine o'clock meeting with the chief, essentially by killing time. I walked around the detective bureau upstairs, reintroducing myself to everyone. I'd been gone from the station for about three years—first working off-site before the indictment, then enduring the entire federal trial mess.

Everyone was over the moon, saying things like, "We're so happy for you," and "We knew this was bullshit," and "We were so scared for you." I felt like a celebrity. It was really great to reconnect with those folks, and I felt a rush of pride returning, something notably absent for close to five years. I never suspected it would be tested again in the next room.

Finally, we sat down with the chief. I signed the papers that restored all my police powers—got my gun back, got my badge back. I was officially a sworn police officer again. But then came the conversation about where I would be assigned, and that's when things took an unexpected turn.

"So, where am I going back to patrol? What happens next?" I asked.

The chief replied, "Well, I'm not ready for that right now. Let's just hold off on this. Let's just stand by. I'm just really not ready for you to go back to work yet."

Did he really just say that? This was the guy who'd said he'd always stand by me. Now this? After everything I'd been through?

Your world gets real small, real fast when expectations don't meet reality. There was a serious conflict in my brain as I tried to process what he'd said. I looked over at the IA lieutenant, and he just looked back at me with an expression that said, "I didn't expect this either, but okay."

The chief was the chief, so we pushed through the meeting. I asked about working some overtime at minimum.

He just repeated, "No, I'm just not ready for that." He turned to the lieutenant. "You know what, Lieutenant, go put him upstairs in the conference room area. There's a special detail unit up there that works three or four days a week for a limited amount of time. Just put him up there. There's an open desk. Just put him up there until I figure out what I want to do with him."

This was not at all what I'd expected, but I tried to roll with it. Maybe things just started back slowly after legal drama like mine. Even so, I didn't like it. The lieutenant walked me upstairs, parked me at a desk, got me a computer, and explained the unusual arrangement.

Somewhat sheepishly, he said, "Hey, you're going to report to the patrol captain, but you're going to be assigned upstairs. This is weird. This has never happened to anybody. I've been here twenty years, never done anything like this with anybody. You're going to report to the captain's secretary for any time off or days off that you need."

I sat there for the remainder of the day, kind of milling around. I had no assignments, literally nothing to do. After lunch, I knocked on Lieutenant Conner's door. "Hey, you got a second?"

"Yeah, man, sure. What's up?"

"What am I supposed to be doing?"

"Nothing, man. Just go sit down with a detective and hang out with them or help them with something. Or just go find something to do." The next thing he said really threw me: "Listen, just go sit down and surf the internet, dude. Nobody's checking on you. You have a free pass, free rein to do whatever you want. Here's a set of keys. Go take one of the cars, whatever you want to do."

This was really weird.

For the first day or two, I played along—surfing the internet, doing some Amazon shopping. But by the end of the week, I was getting restless and ended up talking to the patrol captain, Shu, whom I had worked for in motors previously. He was a pretty straight-up guy and had always been a fan of mine, though he didn't have the backbone when it came to going up and challenging the chief. "I like everything you're doing, man. I'm going to go talk to the chief," he would say, but God bless him, he just didn't have that intestinal fortitude to push further.

"This is really not what I want to do," I told him after four days upstairs. It turned out that the chief had actually made an internal statement that he didn't want me having any public contact, even though I'd been exonerated. We didn't know why, but that was his thinking. I asked the captain, "Will you go back to the chief and fight for me to go work at the training center doing backgrounds in an administrative capacity?"

"Oh yeah, sure, that sounds great. They're behind on backgrounds out there anyway. They're having to farm stuff out to outside companies. This will work out perfectly." Background checks are admin assignments—no uniform, no public contact. In fact, most of what you do in backgrounds is online anyway: phone calls, contacting departments, going to do home interviews of neighbors and business associates, and the like.

But the next morning brought more disappointment: "Yeah, Chad, the chief said no. He's just not ready for you to be out there. He doesn't want you interacting with anybody or being the face of the department. He really doesn't want you interacting with new employees or anything like that."

"Jesus. What the hell does that mean?"

"Chad, the chief's the chief, man. I don't know what to tell you. I'm sorry."

Isolation Takes Its Toll

By the first or second week of February 2019, I was really starting to get demoralized. Every day, people would walk by—well-meaning friends, coworkers, and associates—and ask the same questions:

"Hey, what did you do, man? What happened? I thought you were coming back to work. Weren't you exonerated? Why is he doing this to you? What happened?"

"I don't know. He's just leaving me here. He's not ready for me to go back to work."

This isolation and separation continued for about two and a half weeks. People had their own programs and things going on upstairs, and me wandering around kept them from their work, so I just wasted time on the internet or called my wife fifteen times a day, bugging the hell out of her.

I started getting really anxious because I like to be in perpetual motion; I'm not someone who can just sit there and watch TV all day. I started getting mini anxiety attacks and full-body tingles, with no idea what they were all about. One night, Amiee and I were at the Walgreens by our house, and they had one of those little machines to check blood pressure. Out of curiosity, I sat down and pushed the button, and the reading came up as 164 over 111. My wife is an ER nurse and with wide eyes said, "That's not good. You need to get that addressed ASAP."

Monday morning, I talked to the sergeant who happened to be at the station with the special detail unit. "I have a serious hypertension issue that has come on here in the last few weeks. I didn't have it when I left federal court, but I have it now because of all this stuff I'm going through."

He completed some occupational injury paperwork and sent me straight to the hospital, where they put me on medications. I started trying different blood pressure treatments three times a week at a local medical center. By the middle of March, I was still chomping at the bit. I told the captain and lieutenant that I couldn't just sit there and waste away upstairs in the corner; it was literally driving me crazy.

"All right, well, how about you do some alarm sites?"

Alarm sites are basically when an officer goes out to a 459 ringer (a silent alarm or audible alarm), and it turns out to be bogus. The officer writes a citation and slips it under the door. Your first one's free, but your second one's two hundred dollars, third is four hundred dollars, and up—a penalty to make sure businesses aren't wasting the police department's time. This job was normally handled by a civilian, not a sworn officer.

"The captain's secretary is handling that, but she needs some help, and she's a little backed up. Just go help her do alarm sites."

Oh my God, you gotta be frigging kidding me. Now the dunce cap was permanently fitted to my head. So there I was, cataloging alarm sites and sending invoices to businesses, nothing near what a police officer does. I got out of the office here and there, but not often, and really, this new gig was just a different kind of isolation from my peers. And it got worse. I talked to the chief again, imploring him to let me get back to police work or at least log some overtime, reminding him I'd been exonerated and returned as a sworn police officer. I would soon learn why I had to keep reminding him and why he wouldn't budge. (Ironically, the reason was very similar to the federal judge's.)

I shouldn't have been surprised by the chief's response: "Well, we're going to have another internal affairs investigation."

What? There was already an IA investigation. Everything came back above par; no new information was gleaned in federal court. They've already investigated it, and now they're going to investigate me again?

Why? So the chief could ice me out and avoid dealing with me.

Since Neaderbaomer had bifurcated from Prince and me, on April 8, 2019, Neaderbaomer's trial verdict came back: not guilty. A couple of charges had a hung jury, and the feds ended up just dropping them. He was exonerated, cleared, and coming back to work.

On his first day back, we met up and debriefed about what was going on. I told him about being completely isolated, made to do nothing.

"They can't do that to you," he said. "They can't keep you from working overtime. You're a sworn police officer."

Neaderbaomer ended up having a meeting with the chief, basically telling him that per the California Peace Officer Bill of Rights (POBR), they were violating our rights.

The chief just replied, "Well, I see it differently."

Was this nightmare beginning all over again?

Furniture Mover with a Badge

Might as well throw on another indignity. The police department building, built in the fifties, had tons of asbestos all over the upper floors. The California OSHA had sent an order to do something about it, so the department and city decided to gut the second floor and do a major remodel. They needed to move the entire second floor of the PD building into the basement of the library, two blocks away.

Our second floor held upward of forty offices with desks, huge filing cabinets, chairs, plants, confidential files, and lots of other stuff. During a command staff meeting, a civilian commander said, "Why don't Jensen and Neaderbaomer do the move? If we hire movers, we'll have to pay them and do backgrounds on them because they're handling police files."

When Neaderbaomer and I were approached initially about this subject, Neaderbaomer said, "That's not what we do. We're not here for that."

It was like no one heard our words. Shortly thereafter, Lieutenant Conner pulled us aside: "I just want to let you know the chief really appreciates you guys taking this task on." That really meant, if you don't do this job, you're going to piss the chief off, and then you're never going to go anywhere or be anything here. Kinda sounded like the mafia offering us protection. But what could we do? I'd say our hands were tied, but we needed them to move furniture.

For two months, from the beginning of May through mid-June, Mike and I spent forty hours a week condensing current files, purging old files that had already gone to microfilm, and moving anything that needed to go down to the basement for storage. We must have made a thousand trips in the rickety, musty old elevator, moving chairs, desks, lamps, plants, file cabinets, and any manner of office-related equipment. We then moved everything else from the police station to the library using Mike's personal pickup truck. Again, it was so ludicrous to the point of almost being comical.

Sometime in the middle of May, only for a moment, it looked like things would change. Pomona PD's Lieutenant Steve Congalton returned from the FBI Academy's three-month leadership program. Steve and I were years-long friends; we'd ridden dirt bikes together, gone on family camping trips, and generally hung out. He was a straight-talking guy who always said what he felt.

He saw us moving furniture and pulled me aside in the locker room later. "What are you guys doing?"

"Well, they need this furniture moved. They need people who can be trusted so they don't have to do backgrounds. They just use me and Mike."

"Chad, that's against your POBR rights. They can't work you out of class. They can't make you guys move furniture if you're police officers."

Now, there are certain small tasks that are no problem, like if they ask, "Hey, Chad, can you take that trash out real quick?" Sure. Very menial, small tasks like that are a one-time deal. You do what you've got to do to help the machine move forward. But this was two and a half months of moving furniture every single day. You can't do that.

Other people in the locker room could hear us, and I saw heads poking around the lockers, looking at us, and I thought, *This is going to get back up to the chief, and I still want to get promoted. I don't want to complain out loud and have the chief come down on me.*

I ended up telling Lieutenant Congalton, "I understand what you're saying, and I agree, but I got to do what I got to do; it's not like I have a choice."

The chief got after Lieutenant Congalton for speaking out too. He had IA Lieutenant Conner have a sit-down counseling meeting with him, where he told him to stay in his lane and mind his own business. Like the mafia, if you know what's good for you, you'll shut your mouth. In true mafioso fashion, Congalton was shunned by the chief and punished for his unwillingness to blindly support the chief's orders. I felt bad that he'd gotten beaten up on our behalf, but he'd taken a stand against what was wrong— it was very unusual for someone in the law enforcement hierarchy to actually speak up and try to correct wrongdoings. I won't forget what he tried to do for Mike and me.

This whole thing was a monumental violation of rights. I was pissed and told the IA lieutenant, "This is not Joe's Plumbing. This is a governmental organization, and you have to recognize people's rights. This isn't some willy-nilly little LLC where you can do what you want. There are laws about this."

"Well, the chief sees this differently."

By the middle of June, we had moved basically everything from the sixteen-thousand-square-foot upstairs area of the police station down to the basement of the library, assembled all the cubicles, and put together all the computers. Meanwhile, the sham IA investigation had been going on concurrently.

Ultimately, I had my IA interview with a retired lieutenant from Irvine Police Department who did side jobs as a private investigator. It was probably a three-hour interview, and the guy was, for lack of a better term, being a prick for no reason. I quickly understood that he had prejudged the incident and that I would need to be very patient and articulate. It was tough because I was explaining this incident to a white-collar cop from an uppity Orange County agency, who'd never seen any real street time in a working-class city and didn't understand how policing worked in an environment like the LA County Fair. I could tell it was foreign to him, that he was completely disconnected from that type of experience. I kept my cool until I couldn't take it anymore. I said, "Listen, let's stand up, and I'm going to physically reenact with you exactly what happened." I walked him through it step by step, and the lights went on for this guy after we reenacted it two or three times. Was I finally going to get somewhere?

In the middle of June, I got word that they had completed the third-party IA investigation. The captain called me in, along with Deputy Chief Edwards and Lieutenant Conner. "Chad, this is what we've got. There were three charges that they investigated: excessive force was number one, knowingly omitting factual information from a police report was number two, and unknowingly omitting information from a police report was number three. Charge number one, not sustained. Charge number two, not sustained. Charge number three, they sustained on you."

I said, "I'm sorry, but that says 'unknowingly'—unknowingly omitting information from a police report. Are you serious right now?" They just stared at me. I guess they were being serious. "Well, what's the discipline?"

"We're going to give you two days off."

What? This was absolutely ridiculous, but I was beyond ready to be done with this nightmare, so I replied, "I want to take the days off immediately. I'll take them tomorrow and the next day. I want to put this thing behind me." The IA lieutenant related that Chief Oletti felt he had to give the public, city manager, and city council their "pound of flesh"—they had to get something from this whole mess that had been created by the FBI. I was completely disgusted

Exoneration to Isolation

with these two clowns. As is par for any member of police management, they lost their backbones and their minds as soon as they sold themselves into management and started their journey to executive command staff. But at this point, I was willing to eat a shit sandwich to get this done with because I was sick of having this stuff being held over my head. I wanted to go back to work, and I was tired of having some puppeteer dictating the terms of my life.

About an hour later, the internal affairs lieutenant found me in the basement of the library in the detective area.

"Hey, can I talk to you for a second?" He took me into a closed, empty office.

"I was just talking to the chief right now, and he wants you to Skelly your discipline." A Skelly hearing basically means asking for an in-house hearing with the chief because you disagree with a punishment. He continued, saying the chief felt the Skelly process could be delayed, leaving us to ping-pong back and forth between my attorney and the chief's secretary for eighteen to twenty-four months. By that time, he could retire without having to talk about this to the public. He simply didn't want to deal with it.

"We'll just keep requesting extensions and changes, and you probably won't even end up serving the two days," the lieutenant said.

What the hell? What kind of Mickey Mouse bullshit is this?

However, I was torn. Do you want to bite the hand that feeds you? Do you want to go against the chief when he essentially asks you to use the Skelly system fraudulently?

I walked over to the station and bumped into Neaderbaomer. After I explained to him what had just gone down, he said, "Hard stop. Do not Skelly anything to the chief. You need to get an unbiased third party. Let's Skelly it with the city manager. She knows very little about this whole thing, and you have the right to choose her for the Skelly hearing. She would be the most unbiased party to appeal to. Above all, you're not going to do this ping-pong thing. That's ridiculous. Just so the chief can have his little eighteen to twenty-four months of peace and not have to do his actual job, which is to correct the misinformation and protect the people who have been wrongly accused."

Neaderbaomer spelled out what the chief was attempting to do by skirting the Skelly process. He related that if my IA investigation was completed and signed off and I served the two-day suspension, it would be eligible for a Freedom of Information Act (FOIA) request. But if it were an ongoing Skelly for eighteen to twenty-four months, it wouldn't be eligible for release and therefore couldn't be used against the chief to expose the fact that this was all complete bullshit from top to bottom. If the report and investigation were made public, they would show that the chief, when he was deputy chief, had endorsed and signed off on my original report, the arrest of Carlos, the admin insight, and the subsequent IA investigation. This would be problematic for him because he wanted to distance himself from us and infer that we had done something wrong.

I ended up requesting my Skelly with the city manager, and Schwartz was available for this hearing. A couple of months later, we sat down with the city manager and the city's contracted HR attorney to present the case. The city's attorney brought up the erroneous claim that I'd unknowingly omitted information in my report.

"What information are you alleging that my client didn't put in his report?" asked Schwartz.

"When he was on the stand at state court, he testified that he put his hands up in a stopping motion to the crowd when he was saying 'get back, slow down, stop.' He did not put in his report that he held his hands up."

Schwartz responded, "So you're going to give my client two days off for not putting in his report that he put his hands up? You understand that an officer's report is not evidence. A report is used to refresh someone's memory when testifying in court. That's the purpose of a police report. That's why these reports are not eighty-six pages long. They just note some of the highlights of what happened."

He continued: "So if Chad is a relief sergeant during the daytime, and he kicks back ten reports a day"—a kickback report is when an officer forgets to put a detail in there or forgets to sign it, so it's sent back to them—"you're going to tell me that every single person that gets a kickback is handed two forced days off? That's what you're doing to Chad, because you're alleging that he didn't include a mundane detail in his report that had no bearing on the case whatsoever." The attorney did not have a response to that question; however, the city manager asked if we could explain to her the facts of the federal case and what had happened in the past two years, from the trial to sitting in front of her that day. I was impressed that she was concerned with how and why it had all gone down instead of just going with the popular, media-driven narrative. I felt that she cared, and I was very happy to have selected her for the hearing.

About a month later, I was over at city hall when the HR manager called me in. I was a little concerned, thinking that I was in for disappointment, because the city manager was not in the room, but the HR manager reassured me by saying, "Don't worry—you're going to be happy with this."

She continued, "I spoke at length with the city manager, and she and I agreed that our position on this whole issue is that it is in the past, and we want to move on from it as soon as possible." I had a good relationship with everyone at HR, interacting on various levels with people I'd hired over the years while working in the training division and doing backgrounds. "The city manager was really emotional about this, and the hearing meant something to her. She feels bad for the way you've been treated, especially since you came back here. She's reversing all discipline, and you'll get a letter that says to pay a little closer attention to your report writing; you're a supervisor, and people look up to you."

Fair enough. I signed the appropriate documents, and we were done. Something had finally gone my way—a small bit of vindication.

Back to Patrol—But That's as Far as You're Going

On August 1, I was sent back to day shift patrol, working Monday through Thursday, six in the morning to four in the afternoon as the relief sergeant in charge of the squad. Thank God. I had some semblance of normalcy again, and it felt great to be in stride.

Going into September, we were testing for sergeant again. I had been number one on the list for the last three years and still was. I remember a conversation with Neaderbaomer in the sergeant's office when I told him, "Man, I really, really think the chief is going to do the right thing here. I think he's going to give me a chance, and I think he's going to treat me fairly."

Mike actually laughed out loud. "That guy will never promote you."

"Mike, you don't understand. He's got to promote me. I've tested number one the last four years in a row, and I'm far and away number one. I have been given a green light by the city manager, and she said I was good to go. The chief was even cautioned and instructed by the city manager in his own yearly evaluation last year to start making promotions based on merit."

Mike laughed again. "Chad, stop."

Sure enough, on October 17, two people were promoted to sergeant right in front of me. One was a probationary corporal being promoted to sergeant, and the other was number *seven* on the list. I was heated and hurt—again. Maybe Captain Shu could help.

I explained to the captain how I was feeling and that the entire promotion process was flawed. He felt bad, but you know how it works around here: "Chad, the chief's the chief. It's not you. He just will not promote you."

"How is that fair? To me, that is categorically retaliation."

Again, I tried the IA lieutenant, who had just become Captain Conner. "Can you tell me why a probationary corporal was promoted to sergeant, and number seven on the list was promoted to sergeant, but I can't get promoted? Compare my experience, my resume, and my test scores blind against everyone else on that eligibility list. You will obviously see that I deserved this more than the other six promotable people on that list, and I absolutely beat anybody else far and away."

"Chad, this is the deal. The chief has said to me that he'll only promote you as a hit-and-run promotion when he's leaving. He will not promote you while he's here. He'll only do it while he's on his way out the door." That was when the truth finally got to me in person. The chief was not a man of character, high morals, or standards; rather, he was just a typical politician trying secure as many "likes" and "followers" on Facebook as he could without ruffling any feathers.

At this point, that proverbial light in the back of my head just went dark. There was literally no way out here. I had beaten a federal indictment, gotten through the Skelly process, and finally returned to my home on patrol, and I somehow still had this guy's foot on my neck.

I'm certain that when Prince and I walked out of that courthouse as free men, a deal was being made simultaneously with the FBI that neither of us would ever see another promotion again, even if it was obvious from a merit and tenure standpoint that we deserved it. My theory is that, in exchange for keeping us from advancement, the FBI would back off the steam nozzle on their investigative pressure washing of department policy. This was as good or better than any Hollywood script. I was a free man, yes, but the toll on my career was irrevocable.

I wasn't the only one, and things were about to get very interesting.

A week later, I met Neaderbaomer off-site somewhere. He had also been passed over for promotion several times, and some of our other brethren were peeved about related affronts as well. "I don't know if you know this, but a few other officers and I—lieutenants, sergeants—are filing a lawsuit against the chief and the city of Pomona for workplace retaliation and whistleblower activities, and we're exposing their promotion process here as completely flawed."

"Oh man, I don't know, Mike. That's a big step."

Within that week, a big scuttlebutt drifted around the department that some of the guys were filing a suit, and almost on cue, Captain Conner pulled me aside and said, "You're not thinking about jumping in on this lawsuit, are you?"

"Well, what do you want me to do? I have no recourse here. What can I do?"

"You're going to give up a twenty-year friendship [with the chief] for a ten-thousand-dollar POBR violation?" POBR says that for every violation of your rights, it could be up to a ten-thousand-dollar fine paid to the victim.

I shot right back, "That's pretty rich coming from you, being freshly promoted to captain as a probationary lieutenant! I can't even get promoted to bonehead sergeant. This is ridiculous. I have done everything this department has asked me to do, including moving furniture."

A couple of days later, I got a phone call to meet that same newly promoted captain at a parking structure off-site. I was thinking, *Is this some* Goodfellas *shit? Am I going to get whacked right now?*

When I came rolling up, he was backed into the basement level of the parking structure.

"So the word on the street is you're joining this lawsuit."

"Yeah, I think I am, because again, I have no more recourse. I have nothing. Nobody's in my corner."

"Chad, this is the last time I'm going to tell you. You need to reconsider."

"I'm not going to do that. Hell no. This decision is mine and mine alone." I was legitimately torn. I did have longstanding friendships with the chief and captain, but at some point, you need to stand up for yourself and refuse to take any more shit.

I decided to head over to HR and talk to one of the senior analysts, somebody I worked with quite often. She was in charge of everyone in the law enforcement sector, and we had a very good working relationship.

I knocked on her door. "Can I talk to you for a second? I need some advice. I consider you a friend, and I trust your opinion. This is the decision I have to make: Over here, I have my friendships with the chief and captain, and over here, I have my future, my family, my ability to provide for them, and the guys asking me to join the lawsuit. I'm conflicted because I can't take the abuse anymore, but I want to be promoted, and if I go against the chief, that may never happen."

She got up and closed the door: "You cannot tell anyone that you and I had this conversation because I will get fired for telling you what I'm about to tell you. This is the most classic case of workplace retaliation I've ever seen. This is as textbook as it comes. There is no excuse for what this chief is doing to you. None whatsoever."

I didn't see that coming, but I loved it. At least now I had an unbiased third party telling me the chief was definitely screwing me.

Decision Time

I went home and talked to my wife about it, and we decided to take the leap. The last thing I wanted was more legal bullshit, but they had me backed into a corner. The gloves had to come off.

Our attorneys filed a complaint with the city on a Thursday afternoon, and on Monday morning, the same belligerent captain pulled me aside. "Well, I hope you're happy with what you've done because you've just fucked yourself."

"What are you talking about?"

"You jumped in on that lawsuit. They filed it on Thursday. We all know you just fucked yourself. You should've walked away."

"Because I'm standing up for my rights, I fucked myself?"

He grumbled under his breath and stalked off. After twenty years of being close friends with this guy, that was the last time I talked to him. Standing up for what's right might come at a personal cost, but sometimes it's the only choice you can make.

Within a couple of months of our group filing the complaint with the city manager, the chief abruptly retired. It was obvious from the information that had been obtained through the complaint from our attorney that the chief was going to be put on administrative leave and most likely forced to retire or be released. When he retired, the city appointed the deputy chief to the acting chief position and promoted the patrol captain to deputy chief. It was a relief to get justice for everything the chief had put me through, and I hoped I could finally move forward with my career.

Michael Neaderbaomer, Pomona Police Internal Affairs, (Ret.)

Government conspiracy theories and courtroom corruption? I used to think that stuff didn't really happen. I wanted to believe that, but a steamroller reality check showed me otherwise. When I asked my defense attorney how this could happen when the truth was right in front of us, she said something I'd never heard before: "It has nothing to do with the truth whatsoever. It has to do with an agenda, and whatever fits that agenda is what they make work."

For me, this all started when I received an investigation assignment from the chief's office regarding a citizen's complaint from Mrs. Alvarez. Because she was a dispatcher who worked for Irwindale PD, I prioritized this case and moved it to the front of my workload to get it done as quickly as possible.

During my investigation, I told her over the phone that her claims weren't accurate. She insisted her son hadn't done anything and wasn't involved in anything wrong. I explained, "That's not true. Your son tried to punch the officer. We have video of it."

In April, the FBI called me unannounced and asked, "Hey, we understand there's this video of the kid punching the officer." Keep in mind, they already had the original video from Hobson showing the fair arrest of Alvarez.

I responded, "No, I don't know that to be true." They were alleging this happened, and I told them they were mistaken. In the video, Alvarez *tried* to punch Jensen.

Here's where everything went sideways: Instead of using me as a witness, they made me a suspect.

The FBI claimed that when I told the mother her juvenile son had swung at the officer (which was true based on the first video I saw), I was lying to prevent her from filing her complaint. But here's the kicker—the complaint had already been made. It made no sense, but that was how they had to frame it to fabricate charges against me.

What makes this even crazier is that the first grand jury did not return an indictment. I believe the US Attorney's Office got together with the FBI and essentially asked them to make a case on excessive use of force. If my investigation showed that Chad had done nothing wrong, it was going to defeat the criminal case. But if they also showed a conspiracy at the department, then the department's finding that the use of force was justified would be false and wrong. It would be a juicier story if the department had tried to cover up excessive use of force.

That was how they got the indictment the second time. Here's what went down: The first grand jury had twenty witnesses come in and testify. For the second grand jury, the FBI figured they could falsify what the witnesses had said in the first grand jury and have only one FBI agent testify on what all the other people had previously testified to. This way, they could spin the facts and have just one person control the narrative. The grand jury brought back an indictment within an hour.

How did we get there? I think the FBI agents initially made erroneous assumptions. When our department said the use of force was justified, the FBI didn't understand. I would put any six-month cop against anybody in the FBI when it comes to understanding use of force. The FBI doesn't do police work. They don't answer calls for service, fight with people, or arrest drug dealers. They never have to fight with people under the influence or chase suspects over fences. They just don't have the experience, training, or understanding.

But this is how the FBI operates: They get the indictment first, then shape the evidence afterward. They don't gather evidence, then seek an indictment, because their evidence mining is really good. They figure, "If there's smoke, there's fire." As they went down this path, though, they got in too deep. When the evidence didn't support their theory, they were already committed. FBI agents get kudos if they secure a consent decree against a city, so I suspect that was their ultimate goal. They realized they needed to make it a department-wide conspiracy for their story to make sense.

I pointed out to the FBI that the video had potentially been altered, and I gave them evidence that the video they received was a longer video that had been edited. The FBI tried to fabricate a timeline based on testimony from the suspect's brother and cousin.

According to the government's story, Chad had grabbed the juvenile, the kid had struggled a little, and Chad had swung him around against a fence. I laughed at this because no one puts anybody up against a fence—that's movie stuff. It gives them a point to push off from, and we would never do that.

Here's what actually happened: The officers called for assistance, and they formed a half-circle around the suspects. They were able to handcuff the father and uncle and started walking them to our station about one hundred fifty yards away. The juvenile was following, and Chad wanted some separation between him and the father for safety reasons.

My sister asked me about this incident after seeing the video, so I demonstrated what's called a "twist lock" on her. I said, "Look, I'm not going to move. Do you feel threatened?" She said no.

I explained that the video had captured Chad holding onto Carlos's arm, so I had her hold onto my arm and maintain that position. Then I tensed up, and she instinctively kneed me. When I asked why, she said, "Because you were going to punch me." Exactly my point. When you're holding onto somebody, a video will just show you holding them and them standing still, but in reality, you can feel their entire body and whether they're going to swing, turn, or act in any other violent manner. That's why you maintain body contact.

Chad had felt this kid was going to spin out, which he did. It's not that the kid actually punched him—Chad had felt he was about to get punched. You can see in the video that Chad steps back and tries to protect himself with a forearm strike because he thinks he's about to get hit. Whether he was right or wrong doesn't matter; the question is whether it was reasonable to think so. The answer is yes. Did he respond appropriately? Yes.

Keep in mind that punching someone in the face does more damage and cuts skin open. It's bone-on-bone with your knuckle, and you can potentially ruin your hand, making it impossible to reach your gun or Taser. A forearm strike (not an elbow strike) is actually better for both the suspect and the officer in terms of minimizing injury.

Ultimately, here's how I know the FBI is corrupt: They destroyed evidence. I have a lawsuit against two FBI agents (at the time of this writing). In the suit, even the US attorney has admitted that the FBI agents hid or destroyed evidence, which is called a Brady violation.

But they did everything possible to prevent me from interviewing these FBI agents. We were able to interview the US attorney who'd prosecuted Chad's case, but to every question we asked about probable cause, her response was "I won't answer that question." We literally conducted a deposition with a US attorney who refused to answer questions about whether probable cause existed to bring the indictment, because she knew they didn't have it. They knew 100 percent that probable cause never existed.

It's ironic because the FBI talks about needing transparency, but try making a complaint against an FBI agent. They wouldn't even let me in the building. They don't even have a complaint form. If you want to make a complaint against a cop in California, they have the form available in ten different languages, on the website, in the paper, at the library, at city hall. You can call it in telephonically or make it anonymously. But to make a complaint against an FBI agent? There's no form, and you can't get into the building.

The FBI can't record interviews either. I can't use hearsay as evidence because it's hearsay, but the government can use hearsay against me because I'm supposedly a criminal and a liar. Here's an example: If I said I had a specific conversation with the FBI and it was not recorded, the FBI agent can just get up and say, "No, that's a lie," charge me criminally, and that's it. The perfect example is if an FBI agent is your neighbor, and when they talk about some case without telling you they're investigating it, you mistakenly give them the wrong date for when you had lunch somewhere. Two years later, they can charge you for lying to the FBI, even though it wasn't an active case and you were just mistaken.

That's why the FBI uses this as a leverage tool. It's not about the actual crime—they just get you for lying to the FBI, and they can fabricate evidence because they know that in good police work, witness statements should all be different. If they find any discrepancy, you must be lying, so they charge you with a felony, for which you could spend twenty years in jail.

When it comes right down to it, who do they answer to? There's no one—no one's watching over them, no one's holding them accountable.

That doesn't sit right with me. I'm someone who gives 110 percent. I was in the military, always doing things right, never taking shortcuts. Not in a million years would I ever think I could be indicted, even on fabricated charges. I would never do anything to get me there. The one thing that got me here was that I was in the way of the FBI prosecuting Chad. I was just part of the story that they had to make up.

I think they took me out because they didn't want me testifying in Chad's trial. If Chad's defense team had called me as a witness, it wouldn't have looked good for their case. They could kill three birds with one stone: take me out as a witness, make the story sexier with a conspiracy angle, and help them get a consent decree, which would be a feather in their FBI caps and potentially lead to promotions.

Here's the kicker: The district attorney had seen the video, read the police report, and prosecuted the kid for resisting arrest. So the FBI went after Chad, but the LA County district attorney had agreed that the kid threw a punch based on the same grainy video. I played it over and over, and I just kept thinking, *I kind of see a punch, I kind of don't.* It was questionable—could be more like a shoulder jerk. But the district attorney had prosecuted the kid, and that had not been mentioned anywhere in any FBI report.

The blatant disregard for human beings in the FBI's case and the prosecutor's case is staggering. Chad was facing a forty-year sentence. When you look at people who commit murder or armed bank robberies federally, they get two or three years. I was looking at twenty to thirty years for lying to the FBI.

Think about this and put it in perspective: no injuries. A forearm strike that had resulted in no injuries. On the other hand, online videos, TV shows, and movies show cops doing the wrong thing—beating people with batons, shooting people, driving 110 mph and running them over, tasing them repeatedly, punching people, kicking people in the face, knee dropping people, giving suspects significant and bloody injuries. This Alvarez kid had sustained no injuries from two forearm strikes, and they were going to put a decorated officer away for forty years.

Really think about what the FBI latched onto. It was the story, not the facts. They like the story because it's sexy: "Some juvenile was trying to protect his father." Utter nonsense. Even if they'd assumed it was inappropriate, was this really what the FBI should be worried about, as opposed to significant corruption like cops beating or murdering people? Why were they focusing on this minor issue? To this day, I haven't figured out why they latched onto it so hard, other than the fact that it was a good story for them to get a consent decree.

The way I equate it is this: I've been in the military and involved in incidents that make the news. Tell me a story where you know all the background information, and when it comes out in the news, you'll see that they never get it correct. The story is never what actually happened because they want to spin it a certain way. The full facts of the story never come out. That's consistent with news media across the board, whether liberal or conservative.

That's what the FBI did here, but they went a step further. They'd hidden evidence, destroyed evidence, fabricated evidence, and falsely testified to the grand jury to make it all fit. They had gone so far down that road that their reputations and careers were in jeopardy, so they figured, "Why not do it? We'll get away with it." In Chad's case, they did get away with it. They may not have gotten a prosecution, but they're not being held responsible.

Putting things in proper context is more important than letting the facts fall where they may. That's how I ran my internal affairs investigations. I didn't fabricate facts or take facts out of context. The facts in context are what they are—they're just facts. At the end of the day, you need to let them play out, and everybody's responsible for their own actions.

I'll tell you this: I've been in combat, jumping out of airplanes that were taking ground fire, and I'd rather do that than go through a federal trial again, especially knowing you're innocent. As a cop, you would never think those kinds of things could happen. Now I'm an absolute true believer. I see how the system works.

Obviously, there are bad cops, bad firemen, bad pastors, and bad doctors, and we need to get rid of those people as quickly as possible. But looking at the basic facts, this wasn't one of those cases at all. When you look at what happened, why it happened, and the amount of force that was used, you're left with, "Why are we here?"

CHAPTER TEN

INTO THE SUNSET

Man with a Sword and the Last Day

When we moved to our house up in the high desert in 2007, one of my nearby neighbors worked for the Department of Corrections in Southern California. Bob was a cordial, solid guy, with a big dually pickup and a CWBYBOB license plate.

That first winter, a storm hit us with two feet of snow. We didn't have the equipment to clear our driveway, and the county didn't maintain the road, so we were essentially stuck. Bob had a tractor, and after he chugged on over to clear us out, we got to talking. It turned out we both worked with law enforcement and corrections, and I had friends who work at Ironwood State Prison, so we hit it off.

Then one day, I didn't see him anymore. His truck was there, his motor home, his cars— everything was there. But no Bob. Maybe they moved and sold their stuff? I had no idea, but it seemed odd he wouldn't let me know if they'd moved away. Then one day, my wife and I were out in the driveway when we looked south and saw someone in a wheelchair coming down the road with a lady walking beside him. I asked Amiee, "You know anybody in a wheelchair around here?" She said no. As they got closer, I couldn't believe it; it was Cowboy Bob in that wheelchair, with his wife walking alongside him.

"Hey! What happened?" I called out. "I was wondering where you'd been. Haven't seen you in months."

This was January 2009. Bob worked at a fire camp up in Lake Arrowhead, supervising inmates on a fire crew. He'd done that job for years, driving forty miles from Oak Hills through switchback roads up the back side of the mountain. On one particularly freezing night, he'd left well before sunrise to be at work by 5:30. The roads were icy, and on a tight switchback, his truck slid right off the road and down a one-hundred-foot embankment before coming to rest upside down. He wasn't wearing a seatbelt.

There he was, at 4:30 in the morning, stuck in a ditch while it was well below freezing outside. He quickly realized he had to get out of there, but he couldn't find his phone because it had fallen out of his breast pocket somewhere in the truck. When he tried to get out, he discovered he couldn't move his legs.

He kept telling himself, *If I don't get out of here, I'm going to die.*

He crawled all the way up the side of that mountain. Meanwhile, he was obviously late for work. The inmate workers at his facility were saying, "Hey, Bob is never late. This guy is always on time, by the book, a straight guy. He's never late, ever. Something is wrong."

The inmates convinced the night shift supervisor to call Bob's house and confirm he'd left for work. Sure enough, he'd left at about 4:30 a.m., according to his wife. So they all got in a truck and went to look for him. They drove fifteen minutes down the hill and saw tire marks going off the highway, and when they looked over the edge, there was Bob, about five feet from the roadway. He'd given up trying to reach the top. The inmates picked him up and rushed him to the hospital, which saved his life, but a broken back had left him paralyzed from the waist down.

My mind was blown. Life is about perspective, and the Lord puts people in your life to remind you of that. I thought I had it so bad—I'd just been run through the wringer by the FBI and DOJ, finally gotten back to work, and then had another bomb dropped on me. But recalling Bob's story always humbled me.

I told him how sorry I was, and he looked at me and said, "I want to tell you two things that I want you to learn from what happened to me.

"Number one: put your phone in your front pants pocket or button it into a shirt pocket. Don't leave it loose, because that's your lifeline if something happens to you.

"Number two"—he pointed to my wife—"if she ever asks you to call in sick to work one day and take her shopping, do it. Because I tell you, I worked for the Department of Corrections for twenty-two years, and I never called in sick one day. Not ever. My wife would ask me all the time, 'Oh, honey, take a day, take a mental health day. Call in sick, don't go to work.' But I'd always say, 'I got to go to work, I got to go to work, I got to go to work.' You know what I'm doing right now, Chad? I'm burning all my sick time in this chair."

It was a heavy moment, one that reminded me of my mom telling me to make the best of every situation you have. Like the day we got the flat tire on the way to the courthouse, I could hear her in my head telling me to think of it as, *The other three stayed up.* Unfortunately, that isn't always easy to do, like when your employer dangles the proverbial promotion-carrot in front of you, then exploits it. It was a tough decision to jump into that lawsuit with its potential fall-out, but I had to do what was right for myself and my family.

When I returned to patrol in August of 2019, I never once had anyone from the PD or our community approach me with any type of negative "Hey, you're the guy from the news" connotation; in fact, it was the opposite. Dozens of people thanked me for mentoring their kids, and some of them rode by on their bikes and said, "Hi, Corporal Jensen," along with many other similar moments. It was very rewarding to me, and it was good for my soul to hear that and see that.

By this time, early in 2020, my nemesis, Chief Oletti, had unexpectedly and mysteriously retired, and I was now acting sergeant on the cover shift from eleven in the morning to nine at night. As a patrol sergeant, most of the time, I'm not out there actively taking calls; I just remain available to back somebody up, or I can jump in to help if an officer needs assistance or has a question. So the sergeant stays available 99 percent of the time unless it's a priority zero call—a crime in progress, like a homicide or assault.

I was sitting in my patrol unit at Rebecca and Third Street, backed into a cul-de-sac in an industrial area where I was covered on three sides. You could semi-relax there, as nobody could sneak up behind you. Remember, this was downtown Pomona, not Beverly Hills. Nobody was happy to see you or likely to bring you flowers. I was going over my notes for a catechism class I had that night when a call came in from dispatch.

"Pomona units, we have a report of a man with a sword who just assaulted his brother; victim is bleeding and in need of medical aid at San Antonio and First."

I was on the west side of town, and the situation was going down on the east side. It was crickets on the radio for ten seconds—everyone was waiting for someone else to take it. If you're a patrol officer on some mundane call, you should drop that and go. You can come back to your stale burglary report later. But still, no one spoke up. The newer generation of police officers doesn't seem as motivated or display the ability to multitask as much as the old-schoolers do, so dispatch repeated the call: "Man with sword, medical aid en route."

Finally, I picked up the radio. "Sam Three, Pomona, I'll be en route."

I didn't want to get wrapped up in anything, but the call was far enough away that generally, by the time we get these calls, several minutes have passed. When a victim gets assaulted, they have to get to a safe place, then call the police. Usually, the suspect is long gone by the time we arrive.

I drove straight there—Code 3 (lights and siren) because the suspect was possibly still there and the victim was in immediate danger—stopping or slowing at stop signs and obeying traffic laws. Then they put out another call when I was three or four blocks away: "Suspect is now at San Antonio and Commercial, walking eastbound."

Boom—I was right there. As I pulled up to the liquor store at that corner, I saw a Hispanic male in a white T-shirt with blood running down the back of his head onto his shirt. That was my victim.

As I pulled to a stop to make contact with the victim, a gold minivan approached from the opposite direction. The female driver got out to catch my attention, and it turned out she was the drive-through lady from a local burger joint I'd been going to for twenty years. She was a very nice person, and she always talked to me about problems she was having with her adult son, asking for advice about counseling and what to do.

"Hey, Chad!"

I said, "Hey, stay here, I'll be right back." I had no idea at the time what her connection to this case was; I just recognized her and thought she was saying hello.

I asked the victim, "Where's the bad guy?"

He pointed east, and about two blocks away, I spotted a figure in a black sweatshirt crossing the street, headed southbound.

"Is that him?"

"Yep, that's him."

I told the victim to stay there and radioed dispatch: "Sam Three, Pomona. Your victim's at Second and San Antonio." I stepped back into my police vehicle and sped that way to try to catch the bad guy. Still no word that anyone else was coming to help, so I keyed the mic that the suspect was southbound and I was on my way.

I hauled ass down there and approached the corner, which was occupied by a set of unmanned gas pumps used by trucking services and FedEx. In the middle of that area was a cinder block wall about six feet tall running east-west, with hedges that made an L-turn going southbound. I had lost sight of him momentarily, but in the open gas station area, I spotted him trying to jump the wall. I thought, *Okay, great, I'll just swing around the corner and block him in over there.* I just wanted to contain him and keep him boxed in until other units could get here.

I thought he was jumping over, so I continued around the corner and made a turn northbound. I pulled straight in toward the hedges and saw the bushes moving—he was scrambling through the bushes, trying to get around the wall.

I had my body camera running and my police lights flashing red and blue. I saw him coming through the front bushes, right in front of my unit, so I threw the door open and yelled, "Stop, motherfucker!"

Now, when you work in an inner city community like Pomona, sometimes very direct language is in order. You can't get out there like John Candy at Wally World saying, "Sorry, folks, Wally World is closed." It doesn't work that way. At that point, he had to understand I meant business and wasn't messing around. It's part of command presence—I had to meet any aggression he was bringing and stay one step ahead.

Suddenly, he jumped out of the bushes like Tony the Tiger. Now he was right in front of me with a twenty-six-inch sword, shuffling toward me like a zombie. I was at high ready with my gun and told him, "Let me see your hands. Stop . . . stop!"

As he continued his approach, with my heart racing, I took a couple of steps back, and that was it. At that point, a switch flipped in my brain, and I reverted to my training immediately. The last thing I saw was him bringing his sword up from what I believed was his right side, raising it up katana-style over his head.

Then I fired. I don't have any recollection of squeezing the trigger, no recollection of breathing, counting rounds, nothing. I ended up shooting four times—three in the chest and one in the head. He went straight down to the ground.

A few seconds later, another officer arrived. We did our best with first aid, but the injuries were fatal. Another officer took over, and I stepped back. My faculties started coming back, and I thought, *What the hell just happened?*

These are not conscious decisions you're making in situations like this. When you go to the range, you've got your earmuffs on, and you're thinking about your breathing, sight, trigger, not pushing rounds left or pulling them right. None of that happens here.

Your reaction in a tense confrontation comes from street experience. You don't scream on the radio; you have to control your heart rate, you have to control your respirations because your stomach's in your throat the whole time. But you've got to work through that, and the only way you do that is through muscle memory. That's why in the academy, you go to stress training with instructors screaming and yelling in your face so you can act under pressure and internalize problems so they don't become external problems for you later. In the video, you can see and hear that I stayed calm and in control, but as soon as the captain pulled up and started talking to me, you can hear my respirations become labored.

Because you're running on adrenaline and muscle memory, it's your body and your training making these decisions.

I still get a little emotional about it. One of the captains was out eating lunch when he heard the call and came over to see if I was okay. After things cooled down, they took me to the station, following standard protocol, and separated me from everybody else. Then we proceeded to the briefing room, where you wait for your attorney and/or a support person, usually someone from the officers association. As I was sitting there, I began to think to myself, *Is God trying to tell me something?* What are the odds of me making it through all the federal stuff only to potentially be thrust back into the public eye again? Officer-involved shootings in Los Angeles County are always big stories. I couldn't help but feel like I was going to be put through the spin cycle again.

Neaderbaomer happened to be working that night as watch commander. I asked if he could come sit with me—we'd just been through something crazy together, and I was comfortable with him. But Captain Shu got wind of it and said no. Another blatant violation of my rights. It was my choice who I had as support, and Mike was a member in good standing with the association, not involved in the call whatsoever.

They chased Mike away, and Shu told me to call my wife and let her know I was okay because word spread like wildfire. My wife was working, so I called my oldest son, Blake, also a police officer at our agency. Twenty or thirty minutes later, he showed up at the station.

Shu broke in again: "Your son can't be in here."

"What do you mean he can't be in here?"

The captain stated, "I don't feel comfortable with him being here."

I'm thinking, *What kind of horseshit answer is this?*

My son's a police officer with the city of Pomona and had every right to be there, but they chased him away too. I told the captain I thought he was violating my rights and asked who I should have. He suggested the POA president, who quickly agreed that chasing Neaderbaomer and my son away was absolutely wrong and ridiculous.

Within another couple of hours, two LA County Sheriff's Department detectives arrived. They took my gun, rounds, magazines, body camera—everything. After investigating the scene, they came back and asked if I wanted to watch the video.

I was gun-shy about watching the video after my last experience, when the video hadn't exactly matched what I remembered. I knew I hadn't done anything wrong, but that didn't matter; it still wasn't a good feeling.

The sheriff's guys were cool. They said they were going to play it whether I watched or not. If I wanted to watch, I could. So they loaded it up and hit play. The video showed exactly what had happened, and I was relieved that it agreed with my recollection.

The only thing I didn't like was that the body camera has a fisheye lens that makes things appear a little farther away than they physically are. My recollection was that he'd been right in front of me.

They sent me home and told me to call the chief's office in a couple of days. When I did, they said Internal Affairs was liaising with LA County sheriffs. They also told me I needed to take time off, and they were sending me to talk to a counselor—more standard protocol for every officer-involved shooting.

They sent me to see a mental health guy in the Victorville area, and the counselor turned out to be a retired cop from one of our sister cities, Montclair. He explained that he'd retired from duty and now did counseling for officer-involved shootings. I had an immediate rapport with him. He never made it feel clinical and just talked to me like a normal person.

We talked about everything from the job to what had actually occurred. Oddly, we talked most of the time about how I'd been treated when I'd returned from federal trial and the stress associated with being ostracized from your coworkers. He broke down the causes of the anxiety and stress that I was experiencing and how to work through them since he'd had similar experiences at his old agency. His talks with me were very helpful in processing everything because I was having a hard time sleeping from the totality of what had happened and my treatment when I'd returned to work. Ironically, the shooting itself was not an issue for me. The only part that got to me was that the lady who'd pulled up in the minivan the day of the shooting—the drive-through employee from the burger place—her son had been the suspect. I couldn't believe it when I heard. I had talked to her about her son so many times, only for this to be the final outcome.

Looking back, the only part of the incident that conflicted with me was that I hadn't had an opportunity not to make that decision. The suspect had forced it. Any loss of life is sad. I certainly hadn't wanted to shoot anyone, but I'd had no choice.

Incidentally, the shooting had occurred about fifty yards from his house. He couldn't jump that wall because of the sword at his side, tucked down the inside of his pant leg, so he turned to run through the bushes. He lived only two houses down—we were literally a stone's throw from his front door.

The full story came out later: The suspect ordered that sword off Amazon and walked to the post office with his brother to pick it up. While walking back, I think they stopped somewhere, and he put gray tape on the handle. His brother was walking in front of him in the San Antonio and Holt area when he struck him in the back of the head with the sword, giving him a big six-inch laceration on his skull.

"Why did you do that?" the brother asked, but the suspect took off running, and that was when the victim had called 911.

Aftermath

I had about half a dozen sessions with the counselor, going every couple of days. At about the two-week mark, I was sitting in my garage one night when I got a call from the new chief.

"Hey, Chad, how's it going?"

"Well, I'm working through things."

"Okay, well, you're going to need to report back to work, or you're going to have to start burning your own sick time."

"I beg your pardon?"

"Yeah, we need you back to work, or you're going to have to start using your sick time. I'll see you Monday."

This was completely different from any other officer-involved shooting that had ever been handled by our department. Usually, they'd say, "You let us know when you're ready to come back to work." Never did they call you and say you needed to come back. It was so against everything we supposedly stood for and disgusting that I was getting mistreated again at the hands of these guys.

So I put the uniform on and went ten-eight. This was probably the third week of April, since the shooting happened around March

30. I was back to work for a couple of weeks, still working as the shift sergeant, without extraneous drama, until I got the call about a man with a machete. *What's with all the long-bladed weapons?*

A nut with a machete was banging it on people's cars near the Fairplex (uh-oh) off-ramp from the 10 freeway. Again, the radio was silent; nobody even acknowledged the call, so with no one available, it was up to me. I rolled in that direction, and on the way, I thought, *My luck must be something special—sword and machete calls within a few weeks of each other.* Fortunately, this incident turned out better as the suspect had fled the scene by the time I arrived. I checked the area, turned things over to LA County deputies, and cleared to head back to the station and briefed our watch commander.

Walking out of the watch commander's office, I bumped into Deputy Chief Shu, who always seemed to be full of good news. "Hey, Chad, just so you know, regarding your shooting case, the sword guy's family retained Gaines"—he was the viper attorney from my state trial—"and they're suing the city."

I responded to him, "Okay, well, that doesn't make me happy, but okay. Hopefully it'll be different this time. It's all on video and audio and 100 percent clean. I expect someone will be standing out there on the front steps with full transparency that we don't have any problems with this."

His response was like someone slapping me in the face. "No, c'mon Chad, we can't do that. We can't say anything."

I stared at him for a few seconds, shocked yet again (although I shouldn't have been) over another atrocity happening right in front of me. I asked, "How do the other chiefs and sheriffs do it? Why are they out there every time making statements in support of their officers?"

He just said, "We can't do that," and walked away.

I started shaking a little and got the hell out of the building to the familiar safety of my squad car. But it didn't help. My hands went electric with tingling, my throat throbbed, my chest thumped like a drum, and I got tunnel vision like I was looking through straws.

I thought panic attacks were bullshit, but I was having one right there in the front seat. My academy buddy Bill happened to walk by and noticed something was off.

"Hey, bud, what's up? You okay?" I told him I wasn't feeling well and gave him the quick rundown of what had just happened. He immediately said, "Let's get you up to the hospital. You're fifty years old, and you need to get checked out."

I agreed, we rolled out to Pomona Valley Hospital, where the nurses took me right into a separate area with nobody else around. When they checked my vitals, my blood pressure was something crazy like 165 over 120, and my heart rate was sky high. One nurse stripped off all my gear and told me I needed to calm down.

I did as I was told, slowly feeling better after about half an hour—at least until the deputy chief walked in. *This guy is everywhere.* "Hey, Chad. I just want to make sure this isn't because of something that I said."

Bill and I looked at each other and didn't say a word. I told him I was fine when I really wanted to say, "Just get the hell away from me." He said again that he wanted to check on me and make sure I was okay, but I knew it was bullshit. I turned my attention inward to appreciate the full-circle moment: Bill and I had started our careers together, and now it appeared we had spent our last day on the street together. There we were at the hospital, friends to the end.

That was my last day of work. They sent me home that day, and when I got there, Amiee said, "You're not going back to work. Period. The Lord has told you in no uncertain terms that you're not going to be a cop anymore. It's time to leave, time to retire, time to hang it up. And if you want to go back to being a cop, you're going to be single. I honestly can't take it anymore." I didn't argue. I was tired.

I called Michael Schwartz and told him what happened, and he echoed my wife: "Chad, retire. You beat the federal government. Nobody beats the federal government. Retire, take your win, go to Texas, and live your life. On your headstone, it doesn't mention anything about being a cop—it will say that you were a beloved father, son, and husband below your birth date and the date you die."

He referred me to one of his colleagues who handles workers' comp, because all my heart issues had started when I'd come back to work in 2019 and was put on ice upstairs. After a few months of medical testing and standard procedure stuff with all my other work-related injuries I had sustained over the years, it turned out I had been working with some significant back injuries and heart issues from the stress. Some months later, I got a call from HR around mid-April of 2021, checking on my plans.

I was done.

I'd been plucked out of life and tossed back in so many times. I needed to finally do it on my own terms.

You think big occasions will come with some semblance of fanfare, but it doesn't usually work that way. One moment, I was in my squad car in full uniform, a place I'd been for twenty-four years. The next moment, I was in a hospital bed. And the next one, I was retired. It's crazy how life works. I left a long, proud career, a significant portion of my life, with a small retirement party full of people whom I actually enjoyed working with and admired. No huge department-wide send off with fake fanfare, no big cakes or anyone blowing a kazoo. In a way, it can be bittersweet and a tad disappointing, and it almost always carries some sadness, no matter who you are.

Retiring meant I had to clean out my locker, turn in my gun and badge, and call it a day. On the outside, it was a very mechanical process; inside, it was anything but, and I found myself doing things with a subtle slowness, clinging to the familiar just a little longer. I remember handing my gear over to the equipment officer when he gave me a grin, opened his drawer, and pulled out an ID photo from my first day on the job. We'd had some good times and lots of laughs over the years. I appreciated him making the last day the same.

I went downstairs to my locker and found a pile of memories. There were five or six coffee mugs filled with loose change from back when we actually carried change. All the different tools of the trade I'd collected over the years—a selection of pocketknives, all my old IDs taped to the inside of the locker door, old phone numbers and notes written on the interior walls. Twenty-four years of my life, gone in one day. I finished with my inside locker, turned in the key, and walked outside to my external storage. I opened it up, and there was my old "war bag." Inside was my old city index that listed all the street locations in Pomona and other prominent features so I didn't have to use a map book, old penal codes, vehicle codes, reference materials, and a set of handcuffs that had some history—I'd actually used them to arrest Rodney King one time in the city.

And that was it. It all just goes kaput, and you're off to the next chapter in your life. I remember walking out of there thinking this wasn't the way I'd thought it was going to end. I'd had some really good times and worked with some awesome people. The brothers and sisters I'd worked with on the street made that job what it was. They made it the coolest job in the world.

It was bittersweet, I guess. But I'd had my time, and it was like any other job—you don't work a job for the people you work for, you work the job for the people you work *with*.

Detective Prince Hutchinson, Pomona Police Department

When the FBI initially contacted me, I told them I didn't want to speak without an attorney. After that, I thought everything was fine. I went back to work, life felt normal, and no one was telling me anything. I didn't even have an actual attorney until they got closer to the grand jury proceedings.

What I didn't know was that a grand jury had been convened. I was completely unaware this was happening. I was living my life as usual when I got a call from a sergeant in Internal Affairs asking if I had an attorney. Still having no idea about the grand jury, I was surprised when the chief called and said I needed to go to the Roybal Federal Building in LA.

I asked him, "Am I being charged?" He said yes.

I told my supervisor what was going on, and he said to go home. I didn't know what was happening, what to do, or how to turn myself in. I didn't know if I would be able to post bail or what my charges were.

Once I got home, I tried to explain this to my family. My daughter was only eighteen months old at the time, and I didn't know what was going to happen next. I didn't know if I would be immediately fired or placed on paid administrative leave. This whole situation caught me completely off guard. I still don't fully understand how it reached that level. I don't even think the complainants thought it would get this far. I believe they thought the FBI would investigate and conclude it was no big deal but hoped they might get some financial compensation. However, I think the FBI became so invested that they couldn't walk away.

I also believe some of it was retaliation, particularly against me, because I wouldn't testify against Chad. I refused to testify against him because I hadn't seen him do anything wrong and didn't believe that was the case. The FBI and DOJ were trying to strong-arm me into testifying against him, but I reached a point where I couldn't live with myself if I saved myself by putting an innocent man in prison.

The way the FBI approached me was very interesting. They surveilled me, really thinking they could pressure me into testifying against Chad. But I couldn't just give my own testimony—I would have had to testify to what they wanted me to say to win their case. Before the first trial, they offered me one year of misdemeanor probation to testify against Chad. I told them no. I just didn't believe he deserved what they were trying to do to him.

Here's a fascinating story for perspective: There's a notorious drug dealer in our city whom I had been pursuing prior to being indicted. I had previously arrested him and had been looking for him for a long time. Chad and I were indicted in October 2017, and I ended up arresting this guy again afterward. He had a gun and drugs on him. After a struggle, we got him in custody, and because of his priors, he was facing eight years in prison.

After my indictment, I had a subpoena to testify at his trial. I was dreading it because I was now facing criminal charges myself, and I expected the defense attorney to hammer me on that point. I thought this guy was going to walk. Surprisingly, he took a plea deal and pleaded guilty. I believe he was sentenced to six years but served about three and a half.

Later, in late 2019, after my court case had concluded, I was back at work with our major crimes task force. We happened to pull him over, and he recognized me. He said, "Hey, what's going on, man?" remembering our previous interactions.

I asked him, "Why didn't you take that case to trial?"

He said, "Well, you had a lot going on, and I know you were just doing your job. You didn't deserve to go through all that you went through." He was referring to the FBI investigation and trial. He said I didn't deserve any of the time they were trying to give me. According to the media, we were facing forty years.

For this guy—someone I'd been trying to put in prison, a known drug dealer—to say I didn't deserve to be in prison was extremely telling about this entire situation. This criminal had enough respect and honor toward me that he'd taken a plea deal rather than go to trial because he hadn't wanted me to face the scrutiny of being grilled by his defense attorney while I was facing my own criminal case.

The city didn't care whatsoever. Part of me thinks—and I felt this way prior to that incident—that it's easier to bury you than it is to pay you. That solidified my belief that it would have been easier for them to say, "Hey, they were convicted, and they're in prison, so forget about them. We're all going to move on." When we were not convicted, it created a big headache for them. Now they had to deal with us coming back and the potential backlash—people protesting, community unrest, questions about this and that. It's unfortunate that I feel this way, but when an officer gets into an officer-involved shooting and the officer lives but the suspect is perceived to be unarmed, the department has to go through all this scrutiny. It's much easier for them when the officer dies in the shooting. They get the big parade, the funeral, the sympathies. The family gets the final payout. That's easier to deal with than defending the officer.

This situation made that abundantly clear. It was easier for them to write us off and say, "They were charged, they were convicted, and they're in prison," than it was for them to come out and defend us and say, "They did nothing wrong. They were just doing their job, and this is a malicious prosecution."

That solidified for me that the only person who cares about you is you, and hopefully the person next to you. I committed from day one to the fact that Chad was an innocent man, and whatever happened, happened. I have the mentality of a soldier who's in a room with fellow soldiers when a grenade comes in. Sometimes you have to sacrifice yourself for other people. The fact that the city and department refused to do that for us was very telling of how they view you—expendable.

When we went back to work, they could have just put us back in the positions we were in and moved forward, but they didn't move forward. They moved backward. What is really sad about the entire thing, at least for me, is that I'm not in a physical jail, but since that incident, I've been in a mental jail because this will always be on my shoulders. This will always weigh on me.

CHAPTER ELEVEN

TAY-HAS!

Even before the federal trial, Amiee and I had always said we wanted to retire to Texas. Especially seeing and living through our respective dramas—she worked in an emergency room as an RN and was also well-acquainted with California's bad side—we knew we had to get out of there at some point. And, frankly, we thought you could still be an American in Texas.

We agreed to wait until our daughter Sydney graduated from high school before making the move, but then COVID hit. The kids were all stuck at home doing remote learning, and it was generally horrible for us all. My daughter was midway through her sophomore year when she came to us one day and said, "I know you're retiring soon. If you can get us to Texas before September, I'll be willing to go."

As my retirement approached in May of '21, we put our house on the market, and it sold within two weeks. We had already purchased land in Texas a year prior, so we got ahold of a builder and asked how soon we could build. Unfortunately, this was the middle of COVID, so supply chains were a mess, stalling everything in its tracks. Still, our contractor promised six to eight months without a problem. We were starting in May or June, and he figured we could knock out the exterior work during the dry summer months.

Now we were looking at moving in the middle of June—a month after I retired—but had nowhere to live. I tried to find a rental, but the nearest ones were upward of sixty miles away. I reached out to my realtor, who'd sold us the land. She knew everyone; surely she could help us out. "Hey, Maggie," I said. "Plans have changed. We're not moving out in two years anymore. We're coming in two months. Do you know anyone with a house we could rent for six months while ours gets built?"

She said she'd poke around and see what she could find, but she came up empty, as a lot of people in the area are part-time residents—snowbirds who go north in the summer and stay in Texas here for winter.

One afternoon, Maggie was out having lunch when another realtor named Valerie, who lives in my area now, happened to be sitting at the next table. Maggie mentioned she was looking for a six-month rental for "some good people from California."

Valerie perked up. "Oh, I might know someone. Sherry and Stan live on this gorgeous thirty-acre cattle ranch in Graham—million-dollar property with a big house and swimming pool. They go up to Washington State for the summer, so their place sits empty."

It turned out that Sherry and Stan were actually in town at the time, getting some medical treatment for Stan. Valerie asked Sherry if she'd be interested in having someone house-sit while they were gone. Sherry was immediately interested—until she heard we were from California. "Whoa, whoa, time out. California? Yeah, no. I wouldn't be interested in that."

Maggie let me know the bad news, but I persuaded her to contact Sherry again. Sherry was still frosty, but I explained our situation and asked, "Ma'am, would you at least give me a chance to bring my family out so you can meet us?"

She agreed. "Okay, if you guys want to fly out here, Stan and I will be here this weekend."

So Amiee, the kids, and I booked a flight, grabbed a rental car, and made the drive west from Dallas. We rolled up to a gorgeous thirty-acre cattle ranch on the Brazos River thinking, *Damn, this is really nice.* Sherry answered the door, still apprehensive, but then we got to talking. I told her about my horses (the ranch was horse property too), that I was a police officer, and that Amiee's a nurse—the whole deal.

We ended up talking for three hours. By the end, she was gushing about how much she loved our kids, how well-behaved they were. She finally admitted, "I'll be honest—when I heard California, I thought you'd be a bunch of pretentious jerks. But you're nothing like that."

We ended up signing a lease through December, and our Texas dreams were in full flight.

Meanwhile, we were still in the middle of our civil case against the city of Pomona. We had filed our complaint back in December of 2019, and by now, we were deep into depositions and legal proceedings. The city had hired an outside law firm that was billing them incredible amounts of money (all this public information we obtained through FOIA requests).

By this time, we had built our dream house in Texas and were settling into a new life. Two years into the case, Pomona filed a motion for dismissal in California civil court. Our attorney presented our case, the city presented theirs, and the judge stated to the city, "Not only am I not going to dismiss this case, but I'm advising you that any average juror who hears this case would absolutely rule for the plaintiffs because you have no defense. You need to settle."

Well, the city's attorneys didn't take that advice, choosing to continue with the suit. Fast-forward two more years—they spent over three million dollars defending against our case rather than just settling for what we had asked to be made whole. Ultimately, we ended up settling out of court a week before the jury trial, after five years of extensions and legal maneuvering.

Moving On

It's no surprise I'm grateful every single day to have all that behind me. And moving from California to small-town Texas was a complete and total change. Everything's different here. You can be standing in line at Walmart or the grocery store, and people actually ask you, "Hey, how are you doing?" Coming from California, you're immediately defensive, thinking, *Why are you asking me this?* But people here are genuinely just being friendly. It's funny, too, how everybody knows each other, so you get picked out of the crowd if you're not from here.

For example, we talk fast, California style, while folks in the South and Midwest talk a little bit slower. There were several times my wife and I were at restaurants, trying to blend in by dressing like we're from here, and a waitress would catch us in the act: "So where are you guys from? You're not from here."

It's part of the charm. We absolutely love it out here, the Western lifestyle and Americana of the place.

So how does it all feel now? Do I miss the street? Sure do, once in a while, especially having three of my sons in law enforcement as deputies or police officers. They always have stories or ask for advice, like what to expect on the graveyard shift or SWAT calls.

Yeah, I miss some parts of the job, but it's the people I miss most and the unforgettable experiences and memories with them. To all of you: Thank you. You'll always be in my heart.

Everything has finally settled into how it was meant to be, and it has had as much of an impact on my family as it did on me. My daughter and son are a perfect example. When we came out here, my daughter was just beginning her junior year. She was struggling a little bit in school and socially, but it didn't take long to have a total change of mindset. Now she's in a sorority and enrolled in a nursing program at a college in East Texas—doing fantastic. Our son, who had just graduated from high school one month before our move, had no real direction in life due to COVID forcing him into an "online life." He found direction and purpose and is now thriving, three years deep in the air force overseas. All told, this change of scenery really made a difference in our lives.

The stress level on its own has gone from a million down to ten. I can thank Chevy a lot for that. I still get out and ride at least once a week, most of the time right on our own ten acres or sometimes a nearby trail that leads to a pond and all kinds of peace and quiet. Chevy's my man. He's been here the whole time and brought me the calm and grace I needed at exactly the right time. I often just hang out with him in his stable or the field. I have a perpetual meter in my head that's always moving; anytime my eyes are open and I'm awake, my meter's running. The only time it slowed down during the trial drama was with Chevy. I'd grab a couple of beers, go out there with a handful of carrots, and just watch him eat. It was good therapy then and still is now.

With the positive outcome and closure of the civil case, which wrapped as I was writing this book, I finally feel closure. I never felt like I could really talk to anyone about my experience at Pomona PD. It is cathartic to be in this spiritual and physical place today, where I can release this weight I've been carrying.

Although Chevy was my guy and helped me through the toughest of times, my wife, Amiee, is my true rock. Without her constant steadfast support and love, I'm sure there would have been a different outcome in my life. I am only here where I am because of her ability to counsel me, comfort me, and sometimes be the kick in the ass I needed to continue the fight when I was all but out of gas. The city of Pomona was where we met and will always carry a special place in my heart and our relationship, even after everything.

My story needed to be told, and I'm sure that I'm not the only one. To my fellow peace officers and others tangled in malicious prosecution and federal corruption, don't let 'em win, and remember that the FBI is not your friend.

Be safe out there, and God bless y'all.

ACKNOWLEDGMENTS

First and foremost, I want to acknowledge the absolute power of prayer and the love of our Savior Jesus Christ. The blessings and guidance he provided to me during my life are without limits.

Second, I would like to thank Prince Hutchinson for his rock-solid character, morals, and courage. You are a truly special human being; a lesser person would have given in to the pressure and threats. I am lucky to have gone through this with you.

I also want to acknowledge Michael Neaderbaomer for the incredible strength and discipline that he provided me while going through the whole ordeal. I will never forget the selfless acts you performed that ultimately helped me retain my freedom. I am blessed to have you in my life.

These men personified this quote by Edmund Burke: "The only thing required for the triumph of evil is for good men to do nothing."

ABOUT THE AUTHOR

Chad Jensen proudly served the citizens of the City of Pomona, California as a police officer for twenty-four years. During his tenure, Chad was awarded several awards, including Officer of the Year. He was also recognized for his valuable work within the community he served and was instrumental in training many of the officers working today. As a corporal in 2015, Chad was involved in a low-level use of force arrest that was within policy and the law. In 2017, the FBI began a witch-hunt investigation that resulted in a fraudulent grand jury indictment. This true, firsthand account exposes the corruption, concealment and lies associated with the unfettered and unchecked power within the FBI and the Department of Justice.

ENDNOTES

1. Paul Clinton, "Daryl Gates and the Origins of LAPD SWAT," *Police Magazine*, April 15, 2010, https://www.policemag.com/blogs/swat/blog/15318243/daryl-gates-and-the-origins-of-lapd-swat.

2. United States Code. Title 28, Appendix, Federal Rules of Evidence, Rule 106. *Supplement V, 2000 ed.* https://uscode.house.gov/view.xhtml?req=granuleid:USC-2000-title28a-node248-article1-rule106&num=0&edition=2000.

3. Sandel, W.L., Martaindale, M.H., & Blair, J.P. (2020). A scientific examination of the 21-foot rule. Police Practice and Research. DOI: 10.1080/15614263.2020.1772785

www.ingramcontent.com/pod-product-compliance
Lightning Source LLC
LaVergne TN
LVHW040051080526
838202LV00045B/3587